SECONDARY EDUCATION IN A CHANGING WORLD

Series editors: Barry M. Franklin and Gary McCulloch

Published by Palgrave Macmillan:

The Comprehensive Public High School: Historical Perspectives
By Geoffrey Sherington and Craig Campbell
(2006)

Cyril Norwood and the Ideal of Secondary Education
By Gary McCulloch
(2007)

*The Death of the Comprehensive High School?:
Historical, Contemporary, and Comparative Perspectives*
Edited by Barry M. Franklin and Gary McCulloch
(2007)

The Emergence of Holocaust Education in American Schools
By Thomas D. Fallace
(2008)

*The Standardization of American Schooling:
Linking Secondary and Higher Education, 1870–1910*
By Marc A. VanOverbeke
(2008)

*Education and Social Integration:
Comprehensive Schooling in Europe*
By Susanne Wiborg
(2009)

*Reforming New Zealand Secondary Education:
The Picot Report and the Road to Radical Reform*
By Roger Openshaw
(2009)

*Inciting Change in Secondary English Language Programs:
The Case of Cherry High School*
By Marilee Coles-Ritchie
(2009)

Curriculum, Community, and Urban School Reform
By Barry M. Franklin
(2010)

*Girls' Secondary Education in the Western World:
From the 18th to the 20th Century*
Edited by James C. Albisetti, Joyce Goodman, and Rebecca Rogers
(2010)

*Race-Class Relations and Integration in Secondary Education:
The Case of Miller High*
By Caroline Eick
(2010)

Race-Class Relations and Integration in Secondary Education

The Case of Miller High

Caroline Eick

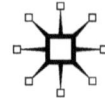

RACE–CLASS RELATIONS AND INTEGRATION IN SECONDARY EDUCATION
Copyright © Caroline Eick, 2010.
Softcover reprint of the hardcover 1st edition 2010 978-0-230-10708-3
All rights reserved.

First published in 2010 by
PALGRAVE MACMILLAN®
in the United States—a division of St. Martin's Press LLC,
175 Fifth Avenue, New York, NY 10010.

Where this book is distributed in the UK, Europe and the rest of the world, this is by Palgrave Macmillan, a division of Macmillan Publishers Limited, registered in England, company number 785998, of Houndmills, Basingstoke, Hampshire RG21 6XS.

Palgrave Macmillan is the global academic imprint of the above companies and has companies and representatives throughout the world.

Palgrave® and Macmillan® are registered trademarks in the United States, the United Kingdom, Europe and other countries.

ISBN 978-1-349-29059-8   ISBN 978-0-230-11442-5 (eBook)
DOI 10.1057/9780230114425
Library of Congress Cataloging-in-Publication Data

Eick, Caroline, 1956–
  Race-class relations and integration in secondary education : the case of Miller High / Caroline Eick.
    p. cm.—(Secondary education in a changing world series)
    1. High school students—Maryland—Baltimore County—Social conditions—Case studies. 2. Students with social disabilities—Maryland—Baltimore County—Case studies. 3. Education, Secondary—Social aspects—Maryland—Baltimore County—Case studies. 4. School integration—Maryland—Baltimore County—History—20th century.
    5. Baltimore County (Md.)—Race relations—History—20th century.
    I. Title.

LC208.4.E35 2010
373.1809752'6—dc22   2010014348

A catalogue record of the book is available from the British Library.

Design by Newgen Imaging Systems (P) Ltd., Chennai, India.

First edition: November 2010

10 9 8 7 6 5 4 3 2 1

To Emmanuel, Francesca, Joseph, and Annie

Contents

List of Tables	ix
Series Editors' Foreword	xi
Preface	xv
Acknowledgments	xix
Introduction	1

Part I The Divided Generation (1950–1969)

1 Memories of Class, Race, and Gender Divisions: Immediate Pre- and Post-Desegregation Years — 17

2 Cautiously Negotiating Social Divides: A Conservative Student Body — 45

Part II The Border-Crossing Generation (1970–1985)

3 Memories of Interracial Peer-Group Affiliations: Integration Years — 61

4 Bridging Social Divides Through Peer-Groups: A Socially Tolerant but Politically Inactive Student Body — 87

Part III The Redivided Generation (1986–2000)

5 Memories of Segregation by Class, Race, Nationality, and Religion: Destabilizing Years of Shifting Demographics — 103

6 Oppositional Self-Segregation: A Student Body
  Sensitized to Discrimination   133

Conclusion   147

Addendum Methodology: The Transparent Historian   157

Notes   175

Bibliography   215

Index   221

Tables

A.1 Pseudonyms arranged by generations, graduation dates,
     and interview dates   167
A.2 Distribution by race, gender, and class (at the time of
     school attendance) of graduates across generations   168
A.3 Population Growth in Miller Town   174

Series Editors' Foreword

Among the educational issues affecting policy makers, public officials, and citizens in modern, democratic, and industrial societies, none has been more contentious than the role of secondary schooling. In establishing the Secondary Education in a Changing World series with Palgrave Macmillan, our intent is to provide a venue for scholars in different national settings to explore critical and controversial issues surrounding secondary education. We envision our series as a place for the airing and resolving of these controversial issues.

More than a century has elapsed since Emile Durkheim argued the importance of studying secondary education as a unity, rather than in relation to the wide range of subjects and the division of pedagogical labor of which it was composed. Only thus, he insisted, would it be possible to have the ends and aims of secondary education constantly in view. The failure to do so accounted for a great deal of difficulty with which secondary education was faced. First, it meant that secondary education was "intellectually disorientated," between "a past which is dying and a future which is still undecided," and as a result "lacks the vigor and vitality which it once possessed" (Durkheim 1938/1977, p. 8). Second, the institutions of secondary education were not understood adequately in relation to their past, which was "the soil which nourished them and gave them their present meaning, and apart from which they cannot be examined without a great deal of impoverishment and distortion" (10). And third, it was difficult for secondary school teachers, who were responsible for putting policy reforms into practice, to understand the nature of the problems and issues that prompted them.

In the early years of the twenty-first century, Durkheim's strictures still have resonance. The intellectual disorientation of secondary education is more evident than ever as it is caught up in successive waves of policy changes. The connections between the present and the past have become increasingly hard to trace and untangle. Moreover, the distance between policy makers, on the one hand, and practitioners on the other has rarely seemed as immense as it is today. The key mission of the current series of books is, in the spirit of Durkheim, to address these underlying dilemmas of secondary education and to play a part in resolving them.

In *Race-Class Relations and Integration in Secondary Education*, Caroline Eick examines student relationships in the years between 1950 and 2000 at Miller High School, a pseudonym for a comprehensive high school in Baltimore County, Maryland. At the heart of her account is the story of how three generations of Miller students interpreted their school experiences and in the process negotiated issues of class, race, gender, and nationality. The events that she describes occurred in the larger context of the struggle in the United States, the South, and Maryland to racially integrate the schools in response to the 1954 Brown v. Board of Education decision.

Hers is an oral history that focuses on the testimonies of graduates during these three generations and considers the extent to which the relationships that were built during their time in school carried over into their adult lives. She identifies these groups as the Divided Generation (1950–1969), the Border-Crossing Generation (1970–1985), and the Redivided Generation (1986–2000). Students, both high school and college, are prominently featured in much of the current scholarship on the civil rights movement and student activism. What distinguishes Eick's work is the fact that she goes further than most studies that explore the political roles of students during the struggles that occurred during these years surrounding desegregation. Her focus is on how students' situated identities, within intersections of class, race, gender, and nationality, were shaped by and in turn shaped institutional norms. It is the story of how students of different backgrounds at different times established relationships that propelled forward the alternative forces of integration and segregation.

In the book, Eick uses the outlook and skills of the oral historian to explore a place that has often been ignored in our histories of school desegregation. Her account places the meanings, beliefs, and perceptions of those participants in the foreground. It recognizes that personal memories are not simply objective descriptions of events. Rather, they are in her words, "...gendered, racialized, and class-based." They remind us that the same events may be seen differently by the various individuals and groups who participate in this or that historical event.

Eick sees her volume as offering an important contrast to another book—Gerald Grant's *The World We Created at Hamilton High* (1988). There is, she notes, an important difference between the two books. Grant's account, as she sees it, is an overly optimistic one that views the less than positive black-white student relationships that prevailed in the school that he studied at the beginning of the movement for integration as improving over time and ending in a period that Grant himself described as constituting "genuine racial equality." Her account, written almost a quarter of a century later than Grant's, tells us a less linear, progressive story. She describes demographic changes that occurred at Miller that brought into the school groups, particularly Russian Jewish immigrants, "angry black city youth," and poorer white youth, who constructed student relationships with a racist discourse. She concludes her book by considering how in different ways the struggle for integration at Miller High both advanced and retarded the ability of these three generations of high school youth to live in the globalized world that was emerging in the last half of the twentieth century and has now fully arrived in the twenty-first.

Eick's book is important in two key respects. It examines the mixed and incomplete results that secondary schools played in the struggle for school integration in the last half of the twentieth century and considers how in the midst of that struggle the American high school served to both unite and divide future generations. Second, her book personalizes the events surrounding one of the most contentious times in our recent history and brings those events into the day-to-day lives of individuals. Events that are typically described at the level of policy and programs become a compelling narrative of real people struggling around real events that affect their lives in deep and profound ways. It is a book that offers a more sober picture of the impact of efforts at desegregation and integration than other similar accounts of this period in our educational history.

Race-Class Relations and Integration in Secondary Education is the eleventh volume to be published in our series. It exemplifies well the combination of social, historical, and comparative approaches to secondary education that we have sought to emphasize throughout, and continues our focus on the interplay between the issues of race and education. As we see the trajectory of the series advancing during the next few years, our intent is to seek additional volumes that bring these issues still further to the attention of studies in secondary education.

BARRY M. FRANKLIN
GARY MCCULLOCH
SERIES CO-EDITORS

References

Durkheim, E. *The Evolution of Educational Thought: Lectures on the Formation and Development of Secondary Education in France.* London: Routledge and Kegan Paul, 1938/1977.

Grant, Gerald. *The World We Created at Hamilton High.* Cambridge: Harvard University Press, 1988.

Preface

The Supreme Court decision in *Brown v. Board of Education*, which struck down public school segregation in 1954, was, for a segment of black and white liberals, a hopeful legislation.[1] Many believed that desegregating public schools would increase economic opportunities for less-advantaged students and attune the more privileged to issues of social justice;[2] and that having practiced meeting across economic, racial, ethnic, and cultural divides during their formative years, generations of students would grow up to be adults who continued to meet across those divides; and that their acquired habits of association would translate into a citizenry who more equitably shared in the political, economic, and social life of its democracy.

However, resistance to school desegregation in the South *and* the North, by whites *and* blacks, before *and* after *Brown*, would challenge what some historians have identified as the power of an educational legislation to serve as a "catalyst for substantial changes in social relations and policies outside of school."[3] Our nation's story of school desegregation would be further complicated by the second-largest wave of immigration within a century in U.S. history.[4] By the 1990s, the meanings attached to categories of racial identity had gained considerable complexity as more diverse peoples entered a country sensitized to issues of civil rights and multiculturalism.[5] Against an imagination of greater equity and opportunity, too many immigrant children of color continued to be funneled into the lower academic tracks of our nation's public schools. Too often they were misdiagnosed as needing special education, and too often trapped in English-language learner classes, out of which they never graduated.[6] Thus at the turn of the twenty-first century, the expectations that might have been attached to *Brown v. Board* by some remain unrealized.[7]

As I write this history, 90 percent of U.S. citizens own a mere 29 percent of the wealth of this nation;[8] schools are more segregated even as the population has grown more diverse;[9] and the number of sexual, ethnic, and religious hate crimes has risen.[10] A cursory glance at our nation's economic disparities and at the racial, ethnic, and recently renewed religious tensions reveals a divided citizenry ill at ease with diversity. Still, we have primarily continued to entrust schools with the task of bringing us closer together, politically and economically. From *Brown v Board of Education* in 1954, to the Elementary and Secondary Education Act of 1965, to the Bilingual Education Act sanctioned by the Supreme Court in 1974, to the Education for All Handicapped Children Act of 1975, and to the No Child Left Behind Act of 2002, we have, over the course of the last half of the twentieth century, continually reappointed schools as our nation's equalizing forces and entrusted them with closing gaps and bridging divides.

Domestically, the story of our pluralistic democracy in the latter part of the twentieth century has been in great part the story of the relationship between our nation's schools and its citizens. Understanding ourselves as diverse peoples occupying shared social spaces at the turn of the twenty-first century is to understand the spaces that we have occupied within desegregated schools over the latter part of the twentieth century; it is to understand the generational transmission of school-acquired habits of association—in sum, to understand school over time from diverse students' points of view.

More than any U.S. institution, public schools have progressively assembled greater numbers of diverse citizens on a daily basis within confined spaces and against their preference or explicit will.[11] In particular, it is within desegregated comprehensive public high schools that the greatest numbers of students at the brink of full citizen participation have continued to meet across economic, racial, ethnic, cultural, national, and religious divides over the past half-century. Patrick Ryan suggests that "more than any other institution, the increasingly comprehensive high schools of the twentieth century redefined the social lives of American youths... [and that] our interpretation of [the comprehensive high school] should be central to our understanding of the country."[12]

Our interpretation of diverse students' relationships with one another, in an institution "as indicative of American society as any institution,"[13] should be central to our understanding of diverse peoples in the United States—particularly at a time in humanity's history when global interdependence is accelerating. How we construct relationships across differences within our own borders, how we bridge or carve divides in our own backyard, will shape how we construct our shared humanity beyond our borders. At a time when we are actively participating in the massification[14] of education across the world, while "we are neither very united nor very comfortable with our diversity,"[15] it is important to take stock of our civic relationships as they have developed since *Brown v. Board*, when we officially attempted to redefine them.

This history explores students' relationships across gender, race, ethnicity, nationality, and class divides as experienced within the architectural, academic, and extracurricular spaces of Miller High, a comprehensive high school in Baltimore County, Maryland, during a half-century marked by desegregation, civil rights movements, suburbanization and urbanization of suburbs, and the second-largest wave of immigration in the United States within a century. It is timely for two main reasons. First, and as previously mentioned, because it contributes to a reflective examination of our national capacity for associative living across census divides since desegregation, at a time when our schools are resegregating at a rapid pace. In this sense, it is a history that sheds light on how one of our important institutions has helped to both unite and divide us—in hopes that "our future may be found in the past's fugitive moments of [harmony] rather than" discord.[16]

This history is timely also because it is written when globalization is redefining the educational aims of nation-states. In addition to the dialectic tension between national democratic forces of cultural homogeneity and heterogeneity,[17] a new tension has emerged at the turn of the twenty-first century between global and local forces. Scholars have begun to suggest that within this new global/local ("glocal")[18] tension, the primary purpose of education—for both developed and developing nations—will be to help citizens adapt to multiple cultural contexts and affiliations within and beyond national borders. This history, then, by bringing into view a recent half-century of student associations in a U.S. school, during a period of great social diversification, offers possibilities for considering the capacities of the U.S. educational system to educate a citizenry at once "American" and global.

My viewpoint, in telling the history of Miller High between 1950 and 2000 from the perspectives of diverse students and as revealed through alumni oral histories, is that generational memories *simultaneously* tell of family, community, the nation, and the world—because ground-level human experiences are many daily interwoven experiences. The experience of school is also the experience of traveling to school, and of having or lacking space to do homework at home after school; of belonging to a church whose pastor is familiar with the school principal, or living in a segregated neighborhood whose elders are rarely seen on school premises; of being brought up Republican or Democrat in a conservative or liberal town; of losing sleep over terrorist threats while still having to attend school. Individual memories, nestled within generational memories and compared across generations, powerfully tell of patterns of intersections between school, social context, and citizens' lived experiences.

I share with the immortal historian Howard Zinn the conviction that "nations are not communities and have never been, [and that] the history of any country, presented as the history of a family, conceals fierce conflicts of interest."[19] Thus I chose to tell the history of one desegregated comprehensive high school from a ground-level perspective—that of students, but more important, from students' situated racial, gender, ethnic, national, and class perspectives. The world is different if one is a poor or wealthy, white or black young woman attending Miller High in the early days of desegregation—different yet from that experienced by a white or black young man, and different still across time periods as gender and race relationships change over time. I also strove not to impose preconceived affiliations between and across racial, gender, ethnic, and class differences as I collected and analyzed graduates' experiences. I strove not to essentialize *agency*, but to look for evidence of it as it came. While using categories such as "race" and "ethnicity" runs the risk of perpetuating social constructs that have no basis in science, they are the categories by which the participants in this oral historical study identified themselves and others; they are the constructs that continue to distinguish the lived experiences of U.S. citizens.

I also share with philosopher Amy Gutmann the conviction that "cultivating mutual respect entails understanding people not merely as abstractions, upon whom [we project our] own conception of what constitutes a good life, but understanding people in their own particularity, with their own lives to lead and their own conceptions of what constitutes a good life."[20] Thus in writing this history, I strove to highlight the particularities of the lives recounted by diverse alumni as they understood them. I hoped to produce a history that cultivated mutual respect even as it examined it.

This book then is a story of U.S. diversity that tells of processes of integration as experienced by three generations of citizens during their formative years at Miller High between 1950 and 2000: the Divided Generation (1950–1960), the Border-Crossing Generation (1970–1985), and the Redivided Generation (1986–2000). It explores our capacities for associative living across differences as these were nurtured or stifled within the classrooms, hallways, and sports fields of Miller High, as well as during dances, extracurricular activities, and across academic tracks. It is at once a history of youth, of diverse students, and of race, gender, and class relations in an institution that desegregated as early as 1956, and of an evolving social context marked by shifting demographics. Over this half-century, the town transformed from rural to suburban to urban-suburban; from a predominantly white middle-class town, alongside which lived a small African American community established in the nineteenth century, to a multicultural population that by the 1990s included Russian immigrants and African American youth newly arrived from city schools.

This work rests on my conviction that the evolving story of U.S. democracy, in particular as it reflects a "world lived in common with others,"[21] is in large part the evolving story of diverse youth in public schools, and that the strength and vitality of a just U.S. democracy rests in great part on how well its citizens learn to cross divides during their formative years.

Acknowledgments

I gratefully acknowledge the following historians of education who have believed in the importance of my work and who have supported my efforts in writing this book. I thank Barbara Finkelstein for generously taking the time to shepherd me through the early stages of my writing, for seeing the published work that I could not then imagine. I thank Eileen Tamura for sustaining my efforts along the way. In her presence, I found strength and resolve. I thank Kim Tolley and Nancy Beadie for articulating, with refreshing clarity, the feedback that always propelled my work forward. I thank my mentor, Philo Hutcheson, for his first-class mind and heart, and the constancy of his integrity and honesty. I thank Sevan Terzian for never hesitating to help when asked, and for providing thoughtful feedback to my early work. I would also like to thank Bob Croninger, sociologist of education and lifelong student of educational policies, who, upon reading the crudest first draft of my manuscript, saw in it a book worth teaching. I thank with all my heart my family—my husband, Joseph, sounding board and wordsmith, who also kept house and home together while I lived the writing, and my son, Emmanuel, and daughter, Francesca, who enthusiastically cheered for me. Last but not least, I thank all the women and men who freely gave their memories—the heart and matter of this book.

Introduction

This is the story told by three generations of graduates who attended Miller High, a Baltimore County comprehensive high school in Maryland, between 1950 and 2000: the Divided Generation (1950–1969), the Border-Crossing Generation (1970–1985), and the Redivided Generation (1986–2000). Each generation of graduates constructed memories of cross-group relationships quite differently, and diverse graduates within each generation told of different limitations and possibilities for negotiating class, race, gender, and nationality lines—if they were women or men, white or black, native-born or immigrant; if they had attended the upper or lower academic tracks, if they had participated in sports, and if they had been of richer or poorer economic background when attending Miller High.

Out of these graduates' recollections emerges a history of integration that extends the story of desegregation beyond struggles to secure equal representation of students across race and ethnicity. It tells of how and which desegregated school spaces translated or failed to translate over time into integrated spaces, and of the forces that sustained and sabotaged *authentic* integration—understood as the "intellectual and social engagement across racial and ethnic groups,"[1] as well as across class and gender divides. Thus this history also adds the dimension of class and gender to the story of desegregation/integration and captures a dynamic account of changing race/class/gender relationships. It is important to underscore that desegregation and integration are distinguished in this history. As suggested above, the former refers to political and legal processes to secure balanced attendance of students across race and ethnicity in schools, while the latter refers to the actual interaction between students across racial, ethnic, as well as socioeconomic groups.

A history of education of student relationships that focuses on processes of integration is timely. First, while the existence of the public desegregated school is presently threatened by the accelerated resegregation and privatization of schools, it continues to hold our best hope for uniting us across differences. By the close of the last century, the notion that authentic U.S. democracy rests on school desegregation had become an idea that more citizens defended than did not. Scholars and the U.S. public in general continued to agree more than they disagreed that "school integration is one of the most urgent educational issues today."[2] However, it remains unclear whether over time desegregated schools have indeed helped to foster authentic integration and if they have helped lessen prejudice. This history is a contribution toward our understanding of the desegregated school as it shapes student relationships, and of the conditions that might foster its capacities as an instrument of integration rather than resegregation.[3]

This history also broadens discussions on desegregation within educational research at large. As one social scientist noted, "[W]hile student achievement data are critical, schools do more than teach academic subject matter: They have a profound potential for shaping individuals and their social networks."[4] In turn, social networks play an equal if not more important role in post-school life opportunities than might achievement. Jencks found that "academic achievement does not have the kind of overwhelming impact on later occupational success that might justify making it the exclusive focus of most research on the outcome of desegregated schooling."[5] Thus a history of education that emerges from the oral historical testimonies of graduates further allows for connections to be made between school relationships of the past and post-school social networks of the present, as graduates reveal whether the interracial and cross-class relationships sealed in school translated into broadened social networks in adult life.

This history is exploratory. It is a beginning. It raises more questions than it answers, and it is certainly not an exhaustive story of all possible configurations of student relationships as they unfolded at Miller High over the latter part of the twentieth century. Nor could such a complete story be told given the many competing historical interpretations of school life and peer relationships. But in important ways, beyond its general contribution to scholarship on student lives—an underdeveloped facet of the history of education—it contributes to scholarship on processes of integration in four important ways. First, it focuses on a suburban high school situated in the orbit of a border city in the border state of Maryland, a barely explored area. Second, it extends and complicates the important story told by Gerald Grant in *The World We Created at Hamilton High*.[6] Third, it offers a framework for historical investigations of processes of integration. Fourth, it foregrounds historical protagonists' meanings, beliefs, and perceptions as historical facts rather than anecdotal evidence to corroborate or challenge established historical facts. As the works of oral historians have suggested, memories are gendered, racialized, and class based. They bring to our attention the importance of examining why different individuals and groups experience the same historical events in very different ways.[7]

The Desegregated Suburban School in Baltimore County

While scholarship has examined civil rights student activism in institutions of higher education, and more recently in high schools, these works do not examine students' unfolding relationships across social categories *within* desegregated schools over time. They tend to focus on students' participations, as black *or* white students, in the political struggles for desegregation and protests for fair treatment.[8] This work examines student relationships within a de facto desegregated school to examine processes of integration over time—a high school situated in Baltimore County, Maryland, in the environs of a border city within a border state.

Baltimore County schools desegregated as early as 1956, for reasons unique to Baltimore and Maryland as a border city and border state, respectively. Maryland's border-state status created a particular context for civil rights movements that contrasted with what Palumbos described as the "triumphalist narrative of struggle, suffering and success," which characterized the Deep South. Maryland activists "began building African American–led institutions to fight racism before a coherent national movement was born...African Americans welcomed white support, and whites generally followed black leadership."[9] Thus Baltimore city schools desegregated immediately in 1954. However, while the city school board adopted a free-choice desegregation policy given the more supportive relationships between African Americans and whites, by the 1960s white flight, later accelerated by ensuing riots following the assassination of Martin Luther King, Jr., rapidly resegregated city schools.[10]

For some third-tier Baltimore County suburbs, those farthest from the city and within rural areas, the story would be different. In the orbit of the city of Baltimore, Baltimore County also desegregated comparatively early, in 1956—but many of its de facto desegregated schools would not resegregate. In 1967, Baltimore County created the "Urban-Rural Demarcation Line," which "preserved exurban areas as farmland in the northern section of Baltimore County."[11] These third-tier suburbs also included the county's long-established black families, whose children, beginning in 1956, would attend desegregated schools. Thus situated in the northwestern corner of Baltimore County, Miller Town is a third-tier suburb that maintained its "rural" status well through the 1970s and early 1980s. At the edge of suburbia and at the edge of white flight, Miller High would remain de facto desegregated—predicated at first on a relatively small number of African American youth, sons and daughters of long-established county families whose roots extended back into the nineteenth century and who were descendents of freed slaves.[12]

By the late 1980s and throughout the 1990s, however, Miller Town and its high school would be drastically transformed from predominantly rural-suburban to urban-suburban, from primarily white and middle class to multiracial and economically diverse. Thus Miller High represents the multiracial suburban high school that has undergone the kinds of demographic changes that many parts of the United States underwent over the latter part of the twentieth century, are presently experiencing and will continue to experience in the foreseeable future. Immigration patterns, particularly since the end of the twentieth century, continue to change the face of public schools beyond urban settings, penetrating suburban schools, which for a while seemed to many middle-class whites as well as middle-class blacks, as we shall see, "safe" from inner-city strife.[13]

Building on Gerald Grant's History

To my knowledge, Grant's work is the only other U.S. educational history that examines student relations within a comprehensive high school in the second half of the twentieth century, and only across race divides. If in Grant's story "things got very bad before they got better, but they did get better," and the school went from "white power to black power to genuine racial equality,"[14] it is in great part, I argue, because Grant's story ends in the early 1980s when the country still rode the wave of civil rights movements, and before the second-largest wave of immigration within a century had flooded U.S. schools. In light of the findings in the case of Miller High, Grant's affirmation of "genuine racial equality" holds only under certain demographic conditions.

Demographics played a crucial role in shaping Miller High students' relationships with school authorities and their interactions with peers across diverse backgrounds. The comfortable interracial and even cross-class relationships (given the middle-class continuum) that students finally achieved during the 1970s were destabilized by an influx of newcomers that brought a new type of white and black person into the community. The new type of white person included the overtly racist and "anti-American" Russian Jewish immigrant youth, and the new type of black person included the overtly angry black city youth whom African American "old timers" accused of corrupting county black youth and addicting them to drugs. The result, during the 1990s, was that a segment of white native youth whom some stereotyped as rednecks, and who might have kept their racist sentiments silent during the 1970s and into the early 1980s, when they were kept in check by integrationist white peers, became visible—as overtly anti-Russian, anti-Jew, and anti-black. By the end of the century, students sported race, ethnic, and national identities in opposition to one another.

But Grant's story also suggests that while black students' political engagement and uprising distanced teachers at first from authentic engagement with students across race, it also eventually forced a school and community reckoning with the race issue. Thus Grant's story suggests that student activism plays an important role in shaping the direction of school relationships. In the case of Miller High, a school situated in a conservative town known to vote overwhelmingly Republican—a town where, as one African American alumna suggested, blacks were "familiar with getting along with white people," and where black and white Miller Town "old timers," as we shall see, preferred the town's status quo relationships for economic as well as social reasons—students remained, across the fifty-year period, politically inactive. During the most daring crossings of race/gender/class divides by students of Miller High, in the 1970s and early 1980s, students never protested against the institutional and institutionalized injustices that they recalled as graduates in their oral testimonies.

Thus informed by Grant's history, I dare to suggest that the lack of Miller High students' political engagement during the 1970s and early 1980s when they were freest to demand changes, given the backdrop of the national civil rights mood and the relatively lax school policies, diminished possibilities for continued integration into the end of the twentieth century. The influx of Russian immigrant youth, many of whom were "anti-American" and racist, and of black youth newly migrated from the city and unaccustomed to desegregated spaces and mistrustful of the white establishment, was greeted by discriminatory institutional norms that had become more rigid by the early 1990s and that exacerbated divisions of race, class, nationality, and even religion. Even as many students in the 1970s and early 1980s had created interracial and cross-class peer groups, as reported by those graduates interviewed, their lack of political engagement against the very sorting mechanisms that they remembered as unjust, coupled with discriminatory institutional practices, allowed for a contingency of black students and poorer students to remain segregated, which would be easily fueled and exacerbated by the addition of racist youth and youth who had not developed habits of association within desegregated spaces.

Thus seismic demographic shifts that brought racist and mistrustful immigrated white and migrated black youth into school spaces, and a lost opportunity for political action by the most democratic generation of Miller High, moved the school not from "white power to black power to genuine racial equality," as was the case of Hamilton High, but from "white power" to "genuine racial equality" to resegregation and aggressive balkanization.

Situated Perspectives and Cross-Group Relationships

Histories that trace processes of integration hold the promise of capturing constructions and reconstructions of categories of social identity as these are shaped within schools, and by extension, to identify more clearly the evolving relative roles played by students, teachers, administrators, and broader social forces in defining the perimeters of relationships across markers of difference. Although using categories such as "race" and "ethnicity" runs the risk of perpetuating social constructs that have no basis in science, they are the constructs that continue to distinguish the lived experiences of U.S. citizens.

The works of educational ethnographers who have been engaged in *multivocal* and *polyphonic* narratives have helped pave the way to explore student encounters across diverse backgrounds.[15] However, lacking the historical dimension that establishes analytic linkages across time periods, ethnographies fail to capture transformations and continuities in students' experiences. Thus this analysis, while not determined by it, is informed by black feminist theorist Patricia Collins' intersectionality theory.[16] The intersectional analytic approach examines the ways in which social markers of difference (race, gender, ethnicity, sexuality, generation, class, religion, and nationality) intersect to shape situated experiences. This approach at once questions the homogenized renditions of group experiences along racial, gender, and other socially constructed categories of identity, and sensitizes the researcher to hierarchies of domination within and across intersecting markers of difference.[17] It further brings to light the shifting nature of group identities. The theory grew out of black feminists' articulation of the limitations of "gender as a single analytical category,"[18] and by extension, the limitations of any single social category to capture the complexity of cross-group relations.

Informed by the theory of intersectionality and committed to an oral historical perspective that foregrounds the subjective experience of history—not to rewrite facts, but to gain a deeper understanding of how and why the same facts are reconstructed differently across historical protagonists—I first identify students' situated experiences within each generation before I analyze more deeply their cross-group relationships. Thus I bring to light the ways that, across time periods, students reinterpret the social categories of race, class, ethnicity, and nationality; the ways they cross and reconstruct those boundaries; and the ways in which they use established categories for new identity purposes.

Historical Protagonists' Beliefs Understood as Historical Facts

In this work, I foreground oral histories as primary sources that tell about meaning, belief, and perception, and not as reflections that corroborate, disprove, or expand on historical events. Alessandro Portelli elegantly captures the possibilities of oral historical inquiry when he writes: "[T]he first thing that makes oral history different is that it tells us less about *events* than about their *meaning*." Portelli boldly asserts that "what informants believe *is* indeed a historical *fact* (that is, the fact that they believe it), as much as what really happened."[19] The intent in this historical analysis was primarily to locate *different perspectives* across intersecting social categories of identities and to capture *meanings* ascribed to relationships in school as historical events in their own right, rather than investigate a single past occurrence. Thus "difference," not "volume," mattered. The goal was not to produce a "collective" memory of Miller High students. Undeniably, such an endeavor would have required "volume," but more important, would have proven futile given the many situated perspectives that compete for the historical meanings of an institution whose population is not homogeneous.[20] While archival research preceded, accompanied, and followed oral historical interviews, and descriptive statistics were derived,[21] the time periods emerged first from narrators' voices; archival research confirmed or corroborated patterns in oral historical testimonies, not the other way around.

Because the focus was on perceptions and meanings ascribed to school relationships by alumni of different backgrounds, the "truth status of individual stories became less important than the value they [were] trying to support in the telling."[22] This work then is also an exploration of the possibilities of historical inquiry to trace evolving perceptions of social groups from multiple perspectives, research having shown that perceptions rather than facts often shape cross-group relationships.[23]

* * *

What This History Is Not

It is important to underscore that this history is not a history of youth subcultures. When youth subcultures come into view, it is because they reflect graduates' experiences of student relationships across categories of race, class, gender, nationality, and/or religion. For example, through a few of the testimonies of the graduates of the 1990s, the punks and goths very briefly come into view. The relevance of punks and goths for this history lies in their being white suburban youth as opposed to an interracial youth subculture. Another example is the prep identity, which is significant for this history because it is the one student identity consistently associated across time periods with upper economic status and superior consumer power. It is a category of identity that by the end of the century included wealthy youth across races and ethnicities. Finally, the jocks, musicians, and other peer groups of the 1970s are significant for this history in that they are the interracial, cross-class, and cross-gender groups that reorganize student relationships toward greater integration and lessened segregation.

The dual focus of this history is to capture students' relationships with peers and teachers from their situated perspectives and to explore the levels of integration across race, class, and other census categories, across time—the comprehensive high school having been assigned the task to bridge the divides of class at its inception, and to bridge the divides of race with *Brown v. Board*, as I explored in some depth in the preface.

Time Period and the Story

The second half of the twentieth century is a particularly fecund epoch within which to investigate youth relationships in high schools because it begins with the greatest expansion of lawfully sanctioned student rights (*Brown v. Board of Education* and *Tinker v. Des Moines*),[24] only to end with almost unprecedented restrictions on students' actions and civil rights (zero-tolerance policies).[25] It is a period that captures a momentous and frenetic cycle in the history of the high school and the lives of its students. High school attendance became a universal expectation, effectively bounding the experience of adolescence with the experience of school, even as the high school itself increasingly struggled for legitimacy.[26] This period has been the era of youth and of the high school, the institution through which all U.S. citizens, since the 1950s, have been expected to live their adolescence.[27]

Also within this arc of time, the real and symbolic power of *Brown v. Board of Education* would be diluted. By the turn of this century, scholarly works began noting a nationwide trend toward segregation and inequality,[28] with some historians tracing it back to the 1970s.[29] Other works began revealing the effect on suburban schools of what has come to be known as the "urbanization of suburbs," whereby more affluent city dwellers move to the suburbs seeking the good life and escaping from poor city schools and crime. The result, as some researchers have pointed out, is that city escapees do not find opportunities for a better future for their progeny, but instead an extension of urban life into the suburb. Then, too, while the Supreme Court decision in *Brown v. Board of Education* originally spoke directly to the black and white divide in the United States, by the end of the twentieth and beginning of the twenty-first century, the landscape of diversity had expanded to include not only people of races beyond black and white, but also people who shared one race but came from very different cultural backgrounds and spoke languages unintelligible to one another. In some schools, more than a hundred different languages could be heard spoken between parents and children. By the end of the century, the meanings attached to racial categories of identity had gained considerable complexity.

Thus not only did I want to trace students' evolving relationships within the high school over almost a half century under *Brown v. Board of Education*, as graduates would reveal these from their situated positions relative to each other and school authorities, but I also wanted to know if subsequent generations of students either transformed or reproduced social prejudice,, and if certain conditions or time periods were more propitious for one or the other. I wondered whether alumni recollections would confirm or disprove what Carnoy and Levin have pointed out, namely that "schools continue to provide Americans with a social experience that is markedly more egalitarian and more open to free choice and possibilities of self-realization than anything that is available to them in the realm of work."[30]

What follows is an account of the evolving nature of students' relationships with each other over the past five decades of the twentieth century as they unfolded for three different groups of Miller High graduates, which I call generations: the Divided Generation (1956–1969), the Border-Crossing Generation (1970–1985); and the Redivided Generation (1985–2000). As mentioned earlier, a complete story of all possible configurations of student relationships as they unfolded at Miller High over the latter part of the twentieth century cannot be told given the heterogeneous student population and the many intersecting situated positions. But the echoed memories of student relationships as reported by the interviewed graduates across race, class, gender, and other divides, as well as memories collected through informal interviews over a two-year period (2003–2005), suggest that enough Miller High graduates perceived cross-group relationships similarly within certain time periods, as well as differently across time periods, to constitute generations. Moreover, groups congealed that articulated in similar language those similarities out of which emerged time periods, which were further corroborated by broader demographic changes in Miller Town. Bound by shared memories of institutional practices, shared demographic configurations, and shared articulations (in use of vocabulary and cultural references) of cross-group relationships, the oral historical testimonies sketched the perimeters of generations. Within each generation, however, nuanced stories emerged from situated perspectives of, for instance, young black middle-class women when compared to those of white middle-class women, working-class men, and so on.

Recollections revealed that authentic integration at Miller High had been the experience of too few students across generations. Interracial friendships that lasted beyond school years were mostly forged during the 1970s and early 1980s, an anomaly in this fifty-year history. A propitious convergence of social forces sustained the Border-Crossing Generation: namely, a broadly shared middle-class status across racial divides; a population of Miller High black students who were mostly children of long-established families in the region and a minority; a student body well accustomed to desegregated institutional settings; and a national consciousness of civil rights that stirred the imagination of an otherwise conservative student population in a conservative town.

Recollections further revealed that cross-class friendships within and across racial divides were rare. For the Divided Generation of the 1950s and 1960s, and for the Redivided Generation of the late 1980s and 1990s, they occurred mostly in upper academic tracks and a few extracurricular activities, most notably in sports. These spaces were the least populated and most insular—spaces for the chosen few. Poorer students' possibilities for participation in upper academic tracks or involvement in after-school extracurricular activities, which proved central to developing relationships with peers across diverse backgrounds, were restricted. Those students of lesser means who could join the upper academic tracks or play in organized sports increased their chances of forging friendships across class and race divides. Thus within elite spaces, encounters across differences were more likely to be friendly. Within the more populous general and vocational tracks, on the other hand, they were more likely to be unfriendly.

Indeed, the general and vocational tracks remained overall divisive spaces, incubators of racial violence. More important, animosities were both reproduced and intensified within these tracks. In the confined spaces of Miller High, students could not avoid each other by dispersing into alternate realities as they might have in the larger community. The cafeteria also remained a place where students continued to segregate as they sought comfort in the familiar and where school authorities consistently left them to fend for themselves. During the 1990s, when demographic shifts brought into school perimeters Russian immigrant and black city youth newly arrived from segregated schools, youth who were mistrustful of the school system and of each other, and who sought refuge in familiar ethnic, racial, religious, and economic relationships,[31] spaces such as the cafeteria only exacerbated students' segregating tendencies.[32]

However, within this overarching account that tells of greater segregation than integration, many nuanced stories complicate and deepen the story. For example, young black and white women in the lower academic tracks in the early days of desegregation of the late 1950s and 1960s experienced more conciliatory racial encounters when compared to those of their male counterparts, and further when compared to those of young black and white women in the lower academic tracks in the 1990s, when they engaged in racial fights. Other stories revealed how an African American youth of means in the vocational track in the early 1960s would be considered "poor" by white peers of lesser means who racialized his economic status. The many nuanced stories strewn across this history testify to the slippery nature of an endeavor such as this one that seeks to build claims on shifting perspectives.

How Miller High students of diverse backgrounds negotiated, organized, and carved relational spaces for themselves, with each other, and with school authorities, and how they integrated or segregated is the story of this history. It is a history that identifies the dynamic interplay between demographics, institutionalized tracking,[33] disciplinary policies, and the power of social class and gender, across time, in loosening or reinforcing students' self-segregating tendencies across racial lines. Many more voices might have recreated a new set of credible experiences that might have expanded or challenged the stories recovered here. But for the graduates interviewed, these were lived experiences, and as such, historic. Then, too, and as suggested earlier, the many nuanced stories of this history resist an authoritative statement about how things were.

Still, I offer claims at the levels of structure and agency. From a structural perspective, I argue that Miller High, born of the progressive era and the social efficiency movement, and which remained effectively unchanged in its architecture and sorting organization since the turn of the twentieth century, could not, by virtue of its classist tracking, create integrated spaces for the majority of its students. Following *Brown*, Miller High black youth entered a high school world of institutionalized classism. Paradoxically, the problem was not so much that Miller High continued to be socially reproductive—since it was also the space within which diverse students, who might never have crossed paths in the broader community, continued to meet—but that it actually exacerbated social tensions for those of lesser means across race, gender, and national identities within its lower tracks, and particularly so under the stresses of major demographic shifts that substantially diversified the student population. Youth of lesser means and/or with lesser prospects of continuing their education into college segregated the most in the lower tracks. Moreover, the greater integration across race and ethnicity within upper tracks was, until the end of the century, predicated on majority white youth representation, suggesting that "integration" within upper-academic tracks was a de facto assimilation of all nonwhite youth and white youth of lesser means into a white middle- to upper-middle-class culture. Further research should examine when integration becomes assimilation rather than the "intellectual and social engagement" of youth across racial, ethnic, class, and gender divides.

Then, too, there is the case of sports. The institutionalized competitiveness of high school sports at Miller High restricted the participation of many students since only those who could compete to win participated. Thus, although sports fields would remain integrated across race and ethnicity, they would also remain elite spaces. More research is needed to examine why so many black Miller High youth were consistently accepted into competitive sports, but so many more talented black youth were never recruited, as this history suggests. Did white coaches apply laxer participation criteria for white youth and more demanding criteria for black youth? Were they striving to keep racial representation tilted toward majority white representation?

From the student agency point of view, I argue that the self-segregating tendencies of Miller High students, when confronted with new social configurations, as was the case for the Divided Generation in the early years of desegregation and for the Redivided Generation of the 1990s, were further exacerbated by distant teachers and punitive disciplinary measures. During the early years of desegregation, many black and poor white youth were met with indifference or outright rejection. During the 1990s, youth fought back and retreated deeper into their racial, ethnic, and national identities under the pressures of zero-tolerance policies.

Broadly and very generally, this analysis suggests that Miller High students of the second half of the twentieth century best bridged social divides in spaces that held high status in the school and that were least populated. It strongly suggests what seems obvious: when young people were respected, dignified, and allowed to explore associations within smaller, more intimate school spaces, they tended to unite rather than divide and to establish friendships across divides that continued into adulthood. Conversely, when young people were considered lowest on the totem pole, ignored, herded into overcrowded spaces, and made to be school visitors through work-release programs, they tended to retreat into groups of familiar backgrounds, to stereotype and fight with those who were different, and to continue to segregate into adulthood.

Parts and Chapters

The analysis that follows proceeds in three parts. Part I, the Divided Generation, is developed in two chapters. In chapter one, "Memories of Class, Gender, and Race Divides," I examine the situated perspectives of Miller High students as young men or women, black or white, richer or poorer—perspectives that proved significant to graduates who attended between 1950 and 1969. In chapter two, I examine the meanings attributed by graduates to the social categories by which they identified their place and that of others at Miller High, and I analyze the spaces, real and imagined, within which Miller High students of the Divided Generation attempted to cross boundaries of race, class, and gender, and within which they cautiously explored alternate identities with peers and teachers.

Part II, the Border-Crossing Generation, is also developed in two chapters. In chapter three, "Memories of Interracial Peer-Group Affiliations," I examine how students reorganized their identities around shared affinities that trumped categories of race, class, and gender, as revealed by graduates who attended between 1970 and 1985. In chapter four, as in chapter two, I examine the meanings attributed by graduates of this generation to the social categories by which they identified their place and that of others at Miller High, and I analyze how they navigated between the various interracial, cross-class, and cross-gender peer groups.

Finally, part III, the Redivided Generation, is also developed following the organization of the previous two parts. In chapter five, "Memories of Segregation by Class, Race, Nationality, and Religion," I examine yet another reorganization of student identities according to dimensions suggested by the chapter's title and as revealed by recollections of those who attended between 1986 and 2000. In chapter six, again as in chapters two and four, I examine the meanings attributed by graduates of this generation to the social categories by which they identified their place and that of others, and I analyze the cultural complexity as well as the rigidity that student segregation acquired by the end of the twentieth century compared with that experienced by the Divided Generation of the 1950s and 1960s.

I follow the body of analysis in the Conclusion, in which I review the social and institutional forces that sabotaged and sustained authentic integration for Miller High students over the second half of the twentieth century, further discuss the claims forwarded earlier, and consider whether indeed Miller High, either within or across periods, offered experiences "markedly more egalitarian and more open to free choice and possibilities of self-realization than anything that [was] available to them in the realm of work."

Finally, in the Addendum, "Methodology: The Transparent Historian," I discuss the voices that could not be recorded and the theoretical and practical challenges presented by *the doing* of oral history.

Part I

The Divided Generation (1950–1969)

Chapter 1

Memories of Class, Race, and Gender Divisions: Immediate Pre- and Post-Desegregation Years (1950–1969)

In the 1950s and 1960s, Miller High[1] students attended a school that had been twice renamed and relocated, and rebuilt several times.[2] Its history extended into the early decades of the twentieth century and was punctuated by landmark investments, one of which in the 1930s transformed Miller Town's older 1914 high school building into the "most modern and best equipped educational plant,"[3] to be modernized yet again in 1960 as the population grew. Throughout its transmutations, it remained a "Main Street" high school, an integral part, both structurally and culturally, of Miller Town, where many of its white middle-class teachers and administrators lived and students could easily cross paths with school authorities on streets, in grocery stores, and in churches. One of the oldest high schools in Baltimore County, Miller High of the 1950s and 1960s was also the only one within a twenty-mile radius, a lone educational establishment in the middle of farm country. One alumnus recalled: "There was a farm behind the high school, and every now and then the cows would get out in the fields...you had to watch out where you were running."[4]

Miller High students in the 1950s inherited an institution with more than one hundred years of history deeply rooted in the white community of Miller Town.[5] This was a high school with long-established rituals. Students both then and today looked forward to "the first day of school...[when] the teacher who had the most tenure rang the bell."[6]

The equally deep roots of Miller Town's African American community, which extend back to the nineteenth century, ran parallel to those of their white neighbors. Generations of black citizens of Miller Town owned small businesses, barbershops, and shoe-repair stores; trained racehorses for wealthy landowners in the area; drove trucks; and provided domestic work for affluent white households.[7] While they lived mostly clustered along Brand Avenue, a road within easy walking distance of the town's high school building, Miller Town's African Americans sent their teenagers on hour-and-a-half-long bus trips to and from the all-black high school in the nearby city of Towson until 1956, when Baltimore County desegregated its public schools.

In the 1950s and 1960s, when "suburbs replaced cities as the fastest-growing residential sector,"[8] Miller Town was still a predominantly rural town and would continue to be so through the 1970s. As one alumnus stated: "We were very middle-class white rural...Oh yeah, we used to get called farmers."[9] Immediately following desegregation, and into the late 1970s, only a handful of African American students would attend Miller High, all of them children of long-established black families of Miller Town whose livelihoods were deeply interconnected with those of their white neighbors and vice versa.

Thus, throughout the 1950s and 1960s, Miller High served an overwhelmingly white population within a predominantly rural setting. Accordingly, until the late 1960s it offered courses in horticulture and animal husbandry, as well as the Future Farmers of America club. It also offered extracurricular activities that reflected the era's broader national politics of concern for international peace—the foreign exchange program and the United Nations Youth Organization—and others such as Students for the American Way, Future Teachers of America, Future Business Leaders of America, and Future Nurses of America, which reflected local demographics, economics, and conservative politics. Miller High provided Miller Town's labor force of farmers, clerks, teachers, factory workers, and businesspeople;[10] educated the college-bound elite; and served as a center for performances, celebrations, fairs, and expositions.[11]

Early Years of Desegregation

When black students began attending Miller High in 1956, they entered a world that had been comfortably inhabited by a majority of white middle-class students and their white middle-class elderly teachers, many who had careers that extended back to the early decades of the century. White middle-class girls in particular, whom teachers and administrators favored, suffused Miller High with their sensibilities and personalities throughout the 1950s and 1960s. To understand Miller High students' differently situated experiences during this early desegregation period, one must first understand the high school world that white middle-class girls participated in creating. It is by contrast to the white female middle-class normative experiences that white male students', black students', and poor students' experiences come into view most dramatically.

White Middle-Class Young Women

Young white middle-class women[12] were at home and at ease at Miller High, an institution that for them held porous boundaries and that they easily entered and exited. These girls moved in a high school world of fluid roles and organizational structures that sustained easy border crossings between school, community, and family life. Student, teacher, and staff could swiftly change to errand girl, grandmotherly figure, or parental authority. As alumna Dorothy Kaufman remembered, permission slips were readily available:

> Mrs. Smith would invariably say, "You want to get out of your study?" Sure. So she'd write a note, and we'd run up to the principal's office and she'd send us to the drugstore to get whatever she needed. I walked right by my father's grocery store and I waived and he waived back... We'd go buy for her and she'd give us money to have a Coke while we were there.[13]

They attended an institution drenched with ancestral memories, where generations of older siblings, mothers and fathers, and even grandparents, lived their high school lives before them. They frequented a high school with a principal, Mr. Lancaster, a "very popular man, both with kids and school,"[14] whom they loved and respected. They lived in a world where their elderly teachers might send their favorite girls on errands to the drugstore in the middle of school days for headache pills.

White middle-class adolescent girls at Miller High in the 1950s and 1960s set the pace and flavor of the school academically, socially, and representationally. Academically, in that they worked harder than the boys, filled the ranks of national honors societies,[15] and never contested assignments, thus raising the bar for achievement. Moreover, they often took care of their male counterparts by offering to do their homework, a habit that would persist into the early 1980s.[16] As one alumnus remembered:

> I had a couple of girls in the classroom that would either take notes for me or sometimes they would give me what they had written as their assignment. And I would copy it or I would change it.[17]

The girls set the tone socially in that they were May queens, homecoming queens, and Miss Senior High; took over the planning of dances; filled the majority of student council seats; and could "almost get away with murder."[18] Alumna Alice Web recalled how easily she duped her teacher by simply being of "respectable background":

> I was looked upon as a good kid, and so I could get away with murder... people whose parents were professionals, or who seemed more respectable [narrator gestures quotation marks as she speaks the word "respectable"]... I would miss the bus on purpose to walk to school 'cause then I'd get there at about ten o'clock and I'd just say, "Oh, Mrs. Reece, I missed the bus again." "Alright," [she'd say] and she'd write me an excuse, you know. And she never said, "You sure miss the bus a lot." You know, I was very rude because nobody stopped me.[19]

Girls also set the tone representationally in that they overwhelmingly determined the content and layout of the yearbook (*The Key*) and forged the opinions of the *Miller Chronicle*.[20] Thus they extolled pride in their school; presented authority figures with reverence; and portrayed students fast at work in classrooms, posing for sports or club group pictures facing the camera straight on and in orderly rows. Linda Moss, a 1969 graduate, captured these young women's involvement:

> We were the doers. We were the people that were on the teams, we were the people that were putting out the yearbook, were the people doing that, you know, class officers... Maybe we were snobs, or elitist, or something like that, but I don't remember feeling that way.[21]

These young women shared an easy transition into a high school where they continued longtime friendships established since elementary school and where authority figures were familiar extensions of parental care.[22] They were also the most likely to receive awards on behalf of the school, such as the then-prestigious Freedoms Foundation Award, "organized in 1949 to further the ideals of American democracy... to create a love of freedom... and to build an understanding of the Constitution and The Bill of Rights."[23]

While in most every way, as alumni reported, white middle-class females made Miller High their home; they were considered smarter than their male counterparts, privileged in the responsibilities awarded them, highly visible in almost every social aspect of the school life, on good terms with teachers, and even allowed to "get away with murder." They differed significantly in what they could do with their high school education and in what a high school diploma meant to them, given the track they attended.

For college-bound middle-class women throughout this period, Miller High was a stepping-stone. Alice Web, graduate of the class of 1954, remembered: "I never questioned that I would go to college...No question I would go. Anybody who had the money could go."[24] Graduates in the 1950s Dorothy Kaufman and Alice Web, and 1960s graduates Sherry Parson, Betty Land, and Linda Moss, alumnae whose fathers could afford college tuition, took it for granted that they would attend institutions of higher education. While Dorothy applied herself and sought good grades, Alice, bored and restless, did minimal work and misbehaved, never doubting that college awaited her.

> If I was interested in the class, then I would pay attention and behave. If there was nothing going on that interested me, I was very apt to get into trouble...There was no pressure to get good grades...never worried about it [about getting into college].[25]

After graduation however, alumnae went on to become teachers. The historically documented trend of women enrolling in teacher programs since the nineteenth century continued throughout the 1950s and 1960s in Miller Town.

For lower middle-class adolescent girls with no allotted budget for college, the academic track was not an option. While they may have excelled academically and widely participated in extracurricular activities, not having money for college meant choosing the next best thing: the commercial track. By default they prepared to enter the business world as secretaries or took art classes.[26]

Whether they attended the academic track to eventually become teachers or the commercial track to be hired as secretaries, looking the part of the middle to upper-middle class was of essence. It was because Judy Law was able to look the part and hide her lower-class status that she could be at home at Miller High. She explained how lucky she had been:

> I was very lucky in one respect, because I was considered to be a cute girl...I was Miss Senior High...My family didn't have any money and they didn't have any status. I was lucky in the respect that I had a grandmother that sewed for women in the Valley. I had gorgeous clothes...My family was very poor and they cared about the image they put out in the community. They wanted me to be clean and well dressed.[27]

Although she knew from the start that college was out of reach, Judy, safely camouflaged in her middle-class look, went about being smart, working hard, and getting involved in a myriad of school activities. She elaborated:

> I was an honor roll student...[in the business track]. My best girlfriend and I were selected [by the principal] to accept an award for our school.

It was called the Freedom's Foundation Award. And we went to Valley Forge with our principal. I have a news clipping of that. I was involved in student council, I was involved in sports, and I was involved in the Future Business Leaders of America. As I got older I was president of that. Because of my background, coming from a family that didn't have money, I was not put on that track [referring to academic track]. I always knew that my parents didn't have the money to send me to college.[28]

In general, high school life for middle-class and middle-class-looking young women was an empowering experience, whether academically or socially. It provided the social and virtual spaces to practice life roles, some with direct correspondence to adult roles that awaited them, and others with no theaters available for them to play out in the immediate and broader cultures of their times. Thus they practiced creating dances to socialize with their male counterparts, actively participated with them in developing "going steady" dating patterns,[29] and immersed themselves in organizing fund-raisers and a multitude of social events. Such social skills, practiced throughout their high school years, could then easily be put to use by these young women as future middle-class housewives. Presiding over the Future Business Leaders of America club or being editor of the *Miller Chronicle,* however, held no equivalent positions in the business communities of the time for these young women. Soon after graduation, some of them felt cheated by such disparities in high school-to-world correspondence, as if they had been lied to.[30] Miller High was the molding ground for a way of life for these young women. They were supported by a network of older middle-class women teachers, who, at the end of their careers, looked the other way when their charges transgressed, often elicited grand-daughterly behavior from them, and spoiled them with privileges and extra money for treats. A network of social players kept things in place for them. As Dorothy explained: "You had all your role models; you had parents, and you had your teachers. If you went to church, you had those people... it sort of gobbled you up, you know, took you in."[31] Judy shared how she "had a lot of encouragement, especially from the older women teachers,"[32] and all alumnae echoed Nora's testimony when she said: "Our teachers took interest in us."[33]

White middle-class and middle-class-looking adolescent girls were at home at Miller High, and made it their home. They saturated its clubs, extracurricular activities, and classrooms with their well-groomed appearances, accompanied by semi-docile, semi-mischievous behavior. They knew the rules well, and knew their teachers' tolerances well, as one knows one's parents' limitations and weaknesses. Their favored position at Miller

High was further corroborated by males' points of view. Alumni Nat Right and Bud Land recalled:

> The girls were treated much better than boys. Girls seemed to get the good side of everything. You had an argument, the boy was always wrong. I don't care what happened. I don't think I'm being prejudiced. That's what they showed me. The girl's always right no matter what.[34]
>
> From a male standpoint, you know, they [faculty and administration] were always easier on the girls.[35]

White middle-class girls navigated the emotional life of school with ease, fluently read the hidden cultural messages of allowable transgressions, and dexterously manipulated the boundaries of acceptable behavior across authority figures and school contexts. Dorothy captured the level of comfort that these young women enjoyed in the presence of authority figures when she fondly recalled when she and her future husband skipped school to see a basketball game, only to meet with Mr. Lancaster, who had stepped out of his principal role to be the ice cream vendor:

> [The] championship basketball game came up between Miller, and I think it was in Kenwood... So, my husband and I were dating at that point and we decided we were going to go to the game... So we took off and there's Mr. Lancaster [the principal] standing there, and he says, "Now, Jim and Dorothy, enjoy the game. I'll see you tomorrow in detention. By the way, I'm selling ice cream." [Doris laughs heartily.][36]

White middle-class girls judged teachers' professional skills apart from their personalities and confidently assessed their abilities regardless of the teachers' dispositions.[37] Whether they judged their teachers as "good" or "bad," these young women could count on them to pay attention to and help them. However, while they shared the world of appearance, comfortable relationships with authority figures, and possibilities for self-expression and leadership roles within Miller High, they did not necessarily share the meanings of their attendance.

For 1950s graduate Judy Law, whose parents were poor, the daily escape from isolation and boredom of country life, opportunity to practice being someone else, and welcoming atmosphere created by teachers and administrators made going to school an experience of deliverance,[38] and the high school diploma itself was the social stamp of approval of her indisputable rights to autonomy.[39] For non-college-bound women like Judy, the high school diploma not only attested to one's official crossing into adult status, but also brought honor to the young adult's family, elevating her in the eyes of the community. Parents whose child graduated from high school were vindicated for past failings and simultaneously acknowledged for the shared success in their daughter's achievements.[40]

For Dorothy and Alice, who were college-bound, high school was a place to have fun on the way to college. While Dorothy worked hard for "two to three hours in the evening, and homework on the weekend,"[41] and assiduously practiced for music recitals and theater plays, she enjoyed every moment of it. This was when she met her future husband, with whom she participated in theater productions. Alice, on the other hand, took advantage of her advantages and got into mischief. She confided: "I'd say come on, we can do it...so we'd sneak off and go swimming, or sneak off and go to the drugstore and get a soda."[42] For 1960s college-bound graduate Sherry, high school was as a "rite of passage that you have to go through," one that she thoroughly enjoyed and during which she "gained good friendships...learned [her] academics [and] to get along with others, leadership skills, those kind of things."[43] For Betty, it was her group of friends that made high school a good experience, and for Linda, it was the support of the faculty and learning that convinced her she could "succeed at stuff."[44]

Whether or not they were serious students, they shared fun times and were at ease with themselves and their environment. These were confident young women for whom "it was a very nice status quo community," a "really a great place."[45] They lived as elite students in the rarified air of the top of the hierarchical totem pole. Dorothy explained:

> In those days...if you didn't go academic, it almost infringed upon your intelligence...People who go to college are it...Of course that's not true, but then.[46]

For these girls, school spirit soared at Miller High,[47] and graduation was "pomp and circumstance," an acknowledgement of their indisputable appurtenance to Miller High.[48]

Poor and Working-Class Young White Women and Men

By contrast, some white girls did not feel "tied to" Miller High. Girls who got pregnant, girls with learning disabilities, and poor girls often dropped out. Indirect descriptions of their lives through others' remembered accounts sketched vague impressions of timid young girls who were lost, alone, and with no place to fit in.[49] Pregnant girls were gossiped about, and whether they disappeared or graduated, they lived a peripheral life in the memories of solidly integrated white girls. Usually more than one girl got pregnant by senior year. The boys who impregnated the girls remained anonymous in narrators' remembered accounts, accounts that well capture the already historically and sociologically explored onus on women of that era for getting pregnant.[50]

If one couldn't dress the part and appear middle-class trim and proper, one was not included. Teachers' support and encouragement, generously lavished on girls who played their parts, were withheld from those who "might not dress well... [and] smelled bad."[51] Familiarity, intimacy, and trust, easily extended to the "right-looking" girls, were denied to the visibly poor. Through Alice Web's middle-class perspective, the very poor came to view in this way:

> The poor kids...who might not...dress well and so on...smelled bad...there was a kind of snobbery about the poor kids. They [the poor] were people who lived in shacks. The teachers didn't try to bring them out...people would talk about poor kids having bugs and that kind of stuff...when they'd do the hair to check the lice, it was always the poor kids who would have the lice, and they would smell funny.[52]

Among those shunned by students and teachers were the poorer white farm boys. Within the predominantly rural community of Miller Town, farm boys represented a significant portion of the student population. However, while Miller High offered agricultural classes and the Future Farmers of America Club, raised its own pigs, and sponsored the yearly fall fair where livestock could be displayed, it was predominantly young men whose fathers were landowners who took advantage of these activities. Boys who would not inherit big farms and for whom training to show prized cattle or choosing the best fertilizers for crops was not useful visited school between jobs. While middle-class boys whose fathers could afford college wore "khakis and...drove their daddy's cars,"[53] James Dean look-alikes like alumnus Robert Heart wore jeans and white undershirts with rolled-up sleeves in which they tucked in their cigarettes, and they drove their bikes to school until they mustered enough money to rebuild old beat-up cars. High school was not a priority for Robert Heart and his friends. They could get through it by just showing up. Robert explained: "I could care less about school...I said, I'm not going to college. So I don't care what I do. I just need to get out, okay?...Just show up."[54]

Robert's recollections of his high school experiences and of those of his friends reveal that the lives of poor farm boys revolved around jobs that made ends meet.[55] There was barely any time to catch up on sleep, let alone do any schoolwork.[56] For these young men, school was a system to endure, retaliate against, reject, feel embarrassed about, or humiliated by. At their best, they suffered teachers; at their worst, they fought them. As Robert put it: "I had a teacher that kicked me out of his class three times...And he had a Volkswagen...And we just picked the Volkswagen [up] and put it on the front steps of the school."[57] Pride gave some a Hollywood-like bravado as they punched out the villain teachers and sometimes walked off into the sunset, vindicated, to become successful entrepreneurs, never to return to the old stomping grounds. Remembering one of his more aggressive friends, Robert said:

> The phys. ed. teacher...kept picking on us...and he screams at us, and you know, take three more. And I said, 'I can't do it. I'm going to work'...So we just walked off the field. So he came over and he got in one of the guy's face, and the guy smacked him and broke his nose, walked right out of the school, and kept on going. He never came back...he is a very successful businessman right now.[58]

These young men studied in the general track and took shop courses. It was in shop class that Robert and many of his friends found a school authority who spoke their language, a male authority figure who simultaneously appealed to their rough edge and nurtured their talents.[59] Many learned early on to fight and fend for themselves. Robert Heart, whose father's death forced his mother to be on welfare for seven years while she cleaned churches, and who remembered coming home from school "and having to take the shoes off and go barefooted because you only had one pair of shoes,"[60] recalled:

> I used to show up at the bus stop, and there was all these seniors...I was always the guy being picked on...they would throw my books over the fence and make me climb over the fence and get them [just before the bus would come]...I was the littlest guy...My mother said to me, "...you have to learn to fight for yourself."[61]

Poor farm boys learned to retaliate when they were picked on—and they were picked on a lot. When peers called them "rednecks and hillbilly,"[62] they "would end up knocking the living hell out of them."[63] Major fights usually happened "on the front lawn of the school."[64] When reflected through the perspectives of those who called some poor farm boys "hillbillies," these were bigoted youth who hated blacks. Sherry Parson remembered:

> Hoo. The fights that went on. There was a bar named Franky's and it was kind of unsavory...and it usually was between a black person and a hillbilly—that's what we called them back then.[65]

African American alumnus Nat Right identified the white student who had knocked his teeth out, as we shall see later in greater detail, as a "hillbilly":

> But honest, when I became an adult, I probably would have hurt him. I thank God our paths did not cross...I forgave him...Not being accepted by the whites or blacks, see, that's where they're coming from. We [the blacks] had a hard time being accepted by the whites. They [the hillbillies] had a double stance. See what I'm saying? The hillbillies were supposed to be the lower-class whites. I can't think of any other way to say it. That was the way it was.[66]

Alumni across time periods alluded to the racism of some of these white boys. They were boys who grew up tough and became sensitized to class differences.[67] Still, although some dropped out—feeling more at ease in the workplace, where they spent most of their time and reaped needed monetary rewards, incentives that found no equivalent match in school—many did graduate. For those who stayed, attendance and diplomas were about beating the system and proving one's endurance. Robert said: "I was determined to be there and to finish the system, even if I had to beat the system somehow."[68] Paradoxically, Miller High was also a place that anchored Robert's life, no matter how unappealing, not unlike a military service.

> I mean, what would I be, who would I be if I didn't go to some kind of school that gave you some regimentation. You know, something that you *had* [narrator's vocal emphasis] to do, something that forced you to do. Someplace that took you to an end.[69]

School was a daily battle not only with some peers, but also with teachers, "teachers who wouldn't bend for [a student]."[70] Unlike the correct-looking white adolescent girls who often wrapped teachers around their fingers, white farm boys felt "picked on" by them.

> They would call on you four times in the class, knowing that you didn't know any of the first questions. And he would continue calling on you for the rest of the day. Trying to make a fool out of you...they were picking on us [farm boys].[71]

For them, teachers' personalities and attitudes toward students were inextricably linked to their efficacy as teachers. While middle-class white girls could have a personal rapport with teachers and still consider their teaching ineffective, working-class farm boys expected a teacher to be at once personable, compassionate, and understanding, as well as be able to deliver differentiated instruction. These young men did not separate a teacher's personal and professional personas.[72] They attended Miller High not as recipients of instruction, but as critical consumers of its deliverance. For them teachers were not familiar people one was to please or receive praise from, but paid agents of a system that owed them.[73]

Other white males of Miller High who were not headed for college, who were not part of the academic track, and who were not necessarily farm boys lived their high school career on the edge of their classroom seats, ready to spring out of them when the local fire department alarm went off. Not one teacher stopped these young men from exiting, sometimes through school windows, when the alarm in town rang. There was a tacit understanding that they would be indispensable to the town in the near future as volunteer firefighters, and therefore should be excused without question from school in the immediate present. Still other white males missed school during hunting season, a yearly ritual inherited from earlier times, and so thoroughly integrated into the cultural expectations of the community that, again, no penalties were exacted for missing school on those days. These were often boys who became the local firemen, mechanics, and construction workers.

Miller High School was not a place that working-class white adolescent men made their own; rather it was a place they visited, a place where middle-class and middle-class-looking white girls ruled, and where in caretaking fashion the girls often did homework for their male counterparts. Robert summed up the white boy–girl divide when he recalled how teachers "expected less of the boys... [and] they weren't going to get anything out of [the boys]"[74]

In general, poorer white male students lived on the edge of Miller High School boundaries. They often missed classes to work, to serve as firefighters in training, or to go hunting with their fathers. While some had it harder than others, they were all expected to find work in the community after graduation. Overwhelmingly, too, they filled the general track; as one alumnus recalled: "[I]t was mostly the guys that were in the general with shop."[75] In the 1950s and 1960s, it was the track considered by students who did not attend it to be a dumping ground for the unable and unwilling student. Bud Land, who had been college bound, described that the general track was meant "for those who were just going to go out and work as laborers. They were the slower kids and the ones that didn't really care about what they were doing."[76] Bud's testimony suggests that middle-class students at Miller High during that period had integrated the notion that those attending the general track were naturally inferior. Ironically, Bud's own memories revealed a young man disinterested in the school curriculum, one who could not wait to get out of Miller High.[77]

White Middle-Class Young Men

Bud Land and his friends were part of a group of white male students who lived through the academic track with ease and nonchalance. Although college bound, they felt no pressure to prove themselves. Bud shared how he "didn't always do what [he] was supposed to."[78] But he knew how far he could deviate from the "conformist" behavior he identified in his white female counterparts, among them his future wife and her friends,[79] before he turned into a seeming "radical." By the very end of the 1960s, a group of white male students, as Bud remembered, were "really out there... they were the ones that were experimenting with some of the drugs... [had] long hair... really long hair... the radical group."[80] These radicals were part of the general track. Just as other alumni recalled, Bud also remembered that tracking determined one's associations "all the time. That was one thing about school. You just moved through with the same group."[81]

Young White Men across Socioeconomic Backgrounds

Bud, a white male college-bound graduate of the late 1960s who put little effort into his studies, shared with Robert Heart, a poor white male graduate of the late 1950s who barely attended school, a perception of teachers as representatives of a system that ultimately did not side with students. Bud recalled: "It was always us versus them—teachers and administrators. I don't remember the teacher getting in trouble for anything. The teacher was the administration, and then there were the students."[82] White males of the 1950s and 1960s, whether rich enough to go to college or not, in the academic track or the general track, as recollections suggested, identified themselves as students *against* teachers. White middle-class girls identified themselves as students *with* teachers. In either case, these young people spoke of their place in school in reference to teachers, and as such, reflected on their place in high school as *students*. Testimonies suggested that white males, such as Robert and Bud, who were poor students regardless of their socioeconomic backgrounds, and who perceived a strong bias on behalf of school authorities in favor of girls, tended to develop more or less antagonistic relationships with teachers. For them, high school was irrelevant to life, and its teachers were out of touch with students.

African American Middle-Class Young Women

For Miller Town black students entering Miller High in 1956, however, school was anything but irrelevant. The first generation of black students who attended Miller High in the first five years or so immediately following Baltimore County's 1956 implementation of desegregation reported absorbing a swell of suspicion, and in some cases, outright animosity. The first years of desegregation at Miller High were particularly difficult for black female students.

In 1956, African American alumnae Annie Cole and Doris Wright entered an institution forced by law to teach them. Unlike their white female counterparts, who experienced an easy transition into high school, Annie and Doris recalled living through a traumatic border crossing, the pains of which scarred their memories. Abruptly uprooted from the all-black school they had been attending, they entered an institution with a foreign tradition and alien history. It was "like being in another world."[83] Annie vividly recalled the first day at Miller High:

> So this particular day...the principal at Washington High said, "Everyone that rides the Miller Town bus from grades seven to eleven, I want you to go to the auditorium," and he said we were to get on the bus and don't ask questions...we did come back to Miller Town and the bus turned into Miller High...so we all went to the auditorium and he [Mr. Lancaster] said, "This is the school you will be attending." My heart went, "Oh, God." I was terrified.[84]

This fearful experience sealed in them longings for the comforts of *their* traditions and a sense of home as they knew it. Annie and Doris remembered missing their old school.

> I missed Washington so much. Oh, boy. The programs we had, and the different things we did, and sometimes, we even could walk and get a hamburger, and do different things. But it wasn't like that here [at Miller High]...It was more stiff [at Miller High].[85]
>
> I think I would have rather graduated from Washington because I had been there from seventh grade.[86]

Unlike the fluid ways in which white adolescent girls traveled Miller High's social and academic boundaries, African American adolescent girls, as Annie reported, were restricted, their behaviors tightly bound to performance in a no-nonsense, businesslike fashion: "learn and behave."[87] To them, Miller High felt strict and harsh, devoid of warmth, emotionally disconnected from the familiar. Its boundaries were rigid and hard to penetrate—so hard that, as Annie's recollection suggested, black moms did not visit:

> My mom used to bring lunches and hot dogs... and she would make cake and bring it to us [at the old school]... it wasn't that kind of atmosphere at Miller High. It was strictly a learning thing. You just had to learn and behave.[88]

Annie walked the hallways fearfully, assailed with spit and unsavory handling by white male students. She spent her energies protecting her body and her feelings, all the while having "to learn and behave." Going to an all-white school took every ounce of effort, took a lot, and finally took too much. Annie dropped out after about a year and a half. She remembered the assaults:

> There was a whole lot of name-calling and spitting on you... They had stairways... you'd go all the way down and you could feel the spit drop on your head and people [boys] would feel your behind, and you know, my mom used to starch and iron my blouse... and it would be ink all over... I couldn't stand it... You know, I left. I graduated from home.[89]

Now and then, however, as Doris recalled, a black girl might be tentatively greeted by a white girl timidly reaching out across the racial divide, more in symbolic gesture than actual friendly contact:

> She [white adolescent female student] used to wave at me [Doris makes the gesture of hand waving timidly with elbow glued to the side of the body], because she was afraid to really talk to me because some of the [white] kids didn't want her to talk to us.[90]

It was within classrooms and during physical education that black teenage girls, as alumni recalled, found safer spaces, moments of intimacies with some teachers, and moments of equanimity with other teachers whose standards for and behaviors toward students remained equal, regardless of color.[91] Subject instruction and teachers' personalities were two separate experiences for Annie and Doris, just as they were for white middle-class or middle-class-looking girls, as discussed earlier. Whether or not a teacher was liked did not determine these students' assessments of the teacher's instructional abilities. However, unlike the experiences of white-middle-class girls for whom neither help nor instruction were ever denied, regardless of the quality of instruction or personalities involved, access to teachers was often denied to black middle-class young women, just as it was to the visibly poor white women. For Annie, relationships with teachers were uneven. Within the engineered racial encounter[92] of the time, and within an all-white school with a long history of being "nicely status quo,"[93] an African American adolescent girl might be denied instruction altogether, not just given poor instruction, because she was black:

> I remember my home economics teacher. She wouldn't help me with my work and I just felt out of place...But would say, "I'm busy now. You'll have to wait until I get to you." But that day never came. If you knew an answer to a question...and if you wanted to know something from the teacher, and you raised your hand to participate in the class, you never got called on. Never.[94]

Doris and Annie moved through an unpredictable, unstable, contradictory, and somewhat surrealistic world inside the walls of Miller High, where peers might assault them on their way to classes or reach out through secret, timid hand signals. Where teachers might support and receive their confidences or withhold instructional assistance.

They were not at home at Miller High, a shifting *Alice in Wonderland* sort of place. Doris and Annie described:

> Going to an all-white school is like being in another world...No, it was just like you were in another world. [Doris shakes her head in disbelief and wonderment].[95]
>
> Some of the teachers understood, and some of them didn't, and some of them just didn't give a damn...just a very uncomfortable situation. It was like everybody had to do this thing because the law said so...And that's the way it was. It wasn't because we're happy. No...it was very hard.[96]

When teachers did nothing to correct and punish white students' verbal or other assaults on them, Doris actively practiced blocking these students' behaviors from her awareness, while Annie sought the help of the principal.

> I don't remember the cafeteria. Some of the things you just blocked out. I don't remember the cafeteria at all...I can do it, I can block out everything around me, and I think I must have been doing it at an early age. Hate to tell you, they'd sit there, talk to me, and I didn't hear. I can block things.[97]
>
> I used to go to the principal. And he used to say to me, "You're doing very well, you're very intelligent." But that wasn't what I was there for, you know..."What you need to do is just try to overcome all of this...It's only a handful of you all, and it's two hundred and some white kids," he said. "It's just going to be hard for me to control this. So what you have to do is just try to overlook it all." So I did for a long while. I couldn't, I just couldn't. I left.[98]

As Annie recalled, the principal, handcuffed by "some of the families that had been in the community helping the school"[99] for years, didn't do much publicly to help change school-wide racist attitudes. Individually, however, he praised black students' work and made public their achievements through honor rolls.[100]

Middle-class African American young women, in the early days of desegregation, unlike their white middle-class female peers who outshined their male counterparts academically and socially, lived in the shadow of their male counterparts, who protected them, chided them, and attracted greater attention. Annie remembered going to her track-star brother for help regarding a bully white boy:

> Sam DiPaglia. I remember him. He used to bother me all the time. He used to punch me in my back, you know... I went and told my brother. My brother said, "Stop whining, don't whine, just ignore him." But he kept on doing it. So I said, "You've got to help me." So he [her brother] went and told him and he [Sam DiPaglia] finally stopped... all I had to do was tell [my brother] who was bothering me and he would go to that person and [that person] would stop... My brother, he got along fine with everybody. They loved him because he was the track star.[101]

For Annie and Doris, Miller High was an imposed exile from the comfortable and familiar, a place where one daily braved humiliations and faced potential bully assaults to the body. They wearily navigated the emotional life of school, spending most of their affective energies in self-protection. Moments spent with a friendly principal, teacher, or peer were far fewer when compared to white middle-class and middle-class-looking girls' easy interactions. None of the black young women were enrolled in the academic track during the early years of desegregation, although they would begin to do so by the mid-1960s.[102] Moreover, none of them, as was the case for many of the poorer white students, participated in after-school activities.[103] Some worked after school, and others were required to head home directly as a measure of protection.[104] While they shared the anxieties wrought by tenuous relationships with authority figures and peers, and together walked a foreign school world that they did not visit after school hours, they differed in how much of the Miller school life they wanted and in their willingness to forge ahead. For Annie, the emotional price exacted for attending Miller High, regardless of her academic achievements, was too high, and the diploma was best acquired in the safety of her home. Thus "dropping out," for Annie, was not a reflection on her poor academic standing, but rather a strategy for protecting her emotional integrity.[105] The energies were spread thin of the black female honor roll student who had "to learn and behave" while her dignity was daily assailed in school, and who was also the babysitter at home and an after-school laborer actively participating in providing financially for her family. Something had to give.[106] For Doris, who did not have to work but who still could not attend after-school activities, there were aspects of Miller High worth exploring should one have had the time. Albeit an alien and at times dangerous world, it seemed to offer possibilities of making her dreams come true and opportunities of which to take advantage:

> Because I went only to eleventh and twelfth grade...As far as I was concerned, we didn't really have enough time at Miller...A lot of the things that were taught there [at Miller High], we didn't have at Washington in Towson [narrator's previous all-black high school]...we were learning something different...I was interested in going into the corporate world...I joined FBLA [Future Business Leaders of America].[107]

For those black young women who graduated throughout the latter years of the 1950s—and nearly all graduated[108]—the graduation itself was as filled with conflicting emotions as had been their attendance at Miller High. A young black girl could be proud of her accomplishments and bring pride to her family yet simultaneously feel disconnected from the school she was graduating from and eager to get out of it. Doris remembered:

> Yes, I was [proud of the achievement]. [But] I just wanted to get out of school. I really wasn't that crazy about school...it was very important, because my parents stressed the fact that you will graduate, you will go to college. But I didn't go to college.[109]

While middle-class African American girls in the early years of desegregation shared feelings of alienation and were denied access to academic tracks like their white female counterparts of lesser means, regardless of their level of participation in the life of Miller High, the experiences of athletic young black males who proved themselves academically were quite different.

Middle-Class African American Young Men

Several African American young men attending Miller High immediately following desegregation were highly esteemed by teachers and peers alike. They were stars on the playing fields in sports competitions and honor roll students in upper academic tracks. Norman Good, a graduate of the class of 1959, was such a young man. For him, Miller High was a place where teachers helped students succeed and peers cheered for one another:

> In many instances I did better in mathematics and English than some of the white students. And they accepted me as one of them when they saw that I could compete.[110]

However, proving oneself at first took some finessing. Norman described how he gingerly offered answers to questions for which he often was the only one to know the answers:

> When there would be questions with regards to grammar, or whatever, and I would know the answer, I would, you know, look around, you understand, being in a situation where I was the only black, I would look around to see if any of my classmates would have the answer, which they didn't. I would shyly raise my hand, and the teacher, my English teacher...a rather elderly lady at that time, she was soon to retire...she would, wondering if I knew the answer or not, she would say, "Norman?" and I would answer it. And this would happen a number of times...then my classmates [would think], "This guy, he knows all the answers."[111]

Norman and young black men like him were the first to set *firsts* for black adolescent students at Miller High and in the county. Norman was the first black student of Miller High to be honored with the Millership Award, and the first black student to win first place in the Baltimore County science fair. Norman and Annie Cole's brother were among the first black students to excel in sports competitions representing Miller High.[112]

Unlike African American adolescent girls who suffered at the hands of bullies and felt "out of place," black adolescent males who competed well academically and in sports felt included by their white peers and never experienced any direct violence.

> In terms of picking on other students, you know, there wasn't. And in terms of violence, there wasn't any violence per se...I can't say any that I personally witnessed. I would hear about it. But I think that the school officials, not the least of which the students, the students of good will, they handled it, handled it well.[113]

Norman's recollections of distant rumors of fights were echoed by many graduates in upper academic tracks within and across time periods, further underscoring the relentlessly insular effect of the upper academic track.

Norman Good, Willie James, and Annie Cole's brother Jimmy Cole[114] were students with reputations as sports stars and scholars. As mentioned earlier, they often played the roles of mediators for younger siblings or other black students whose integration into the social fabric of the school was not so smooth. These young black men participated in after-school activities such as track, high-jumping, basketball, and soccer.[115] Some, like Norman, attributed their scholarly successes to the good training they had received at the all-black high school,[116] and they were out to prove to anyone who might have doubted that they were as good as their white counterparts in the classroom as well as on the field. These were highly motivated students: "I would continue to be the best possible student that I could... I competed in a good-spirited way for grades."[117]

Thus Norman understood his place at Miller High as that of a student there to do the work of a student. Within the rarified air of the upper academic track populated overwhelmingly by white students with means, and on the sports fields after school hours, middle-class black young men forged long-lasting bonds with their white male peers, bonds that extended beyond graduation.[118] Experiencing Miller High from the vantage point of its upper academic track gave black adolescent men opportunities to mingle with whites whose fathers owned stores and whose families held high profiles within the community. This, in turn, augmented their status in the white community.

> I was in a class with an individual whose family owned one of two drugstores. They were prominently known... There was another individual whose dad owned... a jewelry store... and an individual whom my mother worked for doing day work. He made a comment to his mother about how well I dressed and how good a student I was.[119]

For Norman, getting into the upper academic track, however, required much more than good grades on transcripts. Above all, it required firm parental insistence. Immediately following desegregation, black students' transcripts were transferred from the all-black school in Towson, where these students had acquired the confidence of scholarly success built on track records of accomplishments. Norman spoke of his eagerness to prove himself at Miller High and the insistence it took on his parents' part to have him attend the upper academic track:

> I looked forward to the experience [of attending a predominantly white school]. I never doubted my ability... When I arrived at Miller, they had a three-track program... and although they had my transcript of the grades that I had at Washington High, the counselor, I recall, said to my parents, "Well, we think Norman should be able to do very well in the general program." My parents said, "No, we want him in the academic program."... Counselor had some concern about whether I would be able to compete. "Well, okay, fine, what about the commercial track?" [the counselor asked]. My parents said, "No, the academic program." I did very well... I graduated in the top ten of my graduating class.[120]

While counselors were apt to acquiesce to black parents' requests to place their successful students into academic tracks, immediately following desegregation they were neither recommending these students for white higher education institutions nor informing them of options beyond black colleges.[121] Thus graduation held mixed feelings for the college-bound African American young men of Miller High. They had set precedents, brought pride to their families, established collegial relationships with their college-bound white counterparts, but their choices for higher education were restricted, limiting in turn the continuation of their freshly established relationships with "successful" whites. Norman Good's description captures the complexity of sentiments regarding and meanings attributed to graduation and the high school diploma, as he longed to continue his education at a white university:

> Yes. A day of excitement, you know. It was a happy day and in some respects a sad day, because after having been with, at least the ones in the academic program, being with them for three years and knowing that they were going off to universities in different parts of the country, and not knowing whether you would see them...you know, that was the sad part...My classmates, many had gotten scholarships to the University of Delaware, to Towson University, one of them I think the University of Pennsylvania...I would have loved to have gone to one of those schools. Based on my academic record, I would have been able to do so...But anyway, the counselor had just mentioned to me about Morgan State University. I had gotten a scholarship to Morgan, an academic scholarship...I would really have liked to go to University of Maryland.[122]

For other African American male students, attending even historically black colleges was not an option. Middle-class African American young men who did not make it into the academic track moved in more treacherous circles. They were more likely to be attacked and sucked into fights. For them, Miller High was a "black and white" thing.[123] By law, they were desegregated, but in their experience, they were "still segregated."[124] Nat Right, a 1963 graduate, recalled that integration was "something we *had* to do."[125] His recollections sketched one painfully vivid scene immediately following desegregation when Nat was assaulted by a white peer, identified, as mentioned earlier, as a "hillbilly."

> Next thing I know, he hit me in the mouth. I lost six teeth and six had to be pulled. Over the years I have gotten over it. I'll leave it at that. He is dead and gone and my life is changed. I'm a Christian now. He can be forgiven...I was out of school myself for about four months. I had to have gum surgery...That is why I failed that year. Because of the fact I wasn't in school. Because of that.[126]

While college-bound black young men fought for their integration through academic competitiveness, non-college-bound black young men, along with non-college-bound black young women of this early desegregation period, fought for their *bodies* to be allowed and respected within the same high school space that white bodies occupied. For them, fighting was not a rumor but a visceral encounter. Annie's recollection of an infamous Halloween party underscored how black bodies were denied access:

> There was a whole lot of fighting, and I remember this particular time when they had the first Halloween party, they had a dance. And I said, we can go to that now... We almost had a riot... And I didn't understand what was going on, but my brothers did. They said, no, [the white boys] they're trying to start trouble... so it finally escalated. Somebody hit somebody. Somebody called somebody that bad name... And then, that was it... I remember my brother hit somebody in the head with a stick. It just wasn't a good thing.[127]

While black male students enrolled in the upper academic track only heard about skirmishes and thought that school officials handled incidents adequately, other black male students actually took and delivered the blows and found no justice from school authorities, only implorations to try "not to take everything to heart... to ignore most of it."[128] Similarly, poorer white male adolescents, who did not participate in sports and were not enrolled in the upper academic track, were more likely to find themselves embroiled in fights, but they also directed their aggressions toward male teachers.

Thus the lower general track, attended mostly by males, primarily poorer white males (many of whom, as earlier suggested, were racist farm boys) and black middle- and working-class males, became the site of racial strife. All the while, Mr. Lancaster, judged the fairest of principals by whites and blacks alike and beloved by his community, felt powerless to provide any redress for the racist violence against his black students whom privately he praised. However, by the mid-1960s, life at Miller High became more bearable for young black women and men, as graduates' recollections suggested.

Settling into Desegregation

For Dotty Moris, the first day of high school was a far cry from the experiences of Doris Right and Annie Cole in the early days of desegregation. Dotty remembered her first day in 1963 as uneventful: "But as I can remember, it was just a normal day. We walked. So it was quite a few of us that walked together...and then we met with others, the same kids that went to Miller Elementary."[129] Dotty began her high school career alongside neighborhood friends who had graduated with her from primary school. Following an easy transition into a new level of schooling, she soon expanded her circle of friends across racial boundaries:

> And then I met other people, new friends...we just became best friends. So you had an opportunity to meet new people that you just didn't even know that were in your area...I met a lot of white friends.[130]

Dotty had "a good relationship"[131] with her teachers and got involved in many school activities early on, pursing interests that she had begun exploring in middle school.[132] In Dotty's story, things were no longer just "black and white" as they had been for Annie and Doris. Dotty, who was not spending her energies protecting herself from physical and emotional assaults and who had already practiced being a student in a desegregated elementary school, remembered relationships with peers less in terms of skin pigmentation and more in terms of friendly or unfriendly behaviors across races.

As Dotty stated, while there were white students who still gave black girls the cold shoulder, there were outright bully black students one strove to avoid.[133] *Bad* girls and boys, according to Dotty, were homebred, and student behavior was a direct reflection of familial upbringing, whether white or black. The black and white polarization of the early desegregation period was further challenged in the mid to late 1960s with the beginnings of interracial couplings at Miller High.[134] Less than a decade after the county had implemented desegregation, and within fewer than two generations of graduates since the Halloween dance incident, a lone interracial couple in a still very rural Miller Town dated and married. Their relationship forged within the world of high school.

For Dotty, Miller High in the 1960s was a great school that upheld high standards of achievement and excelled in county sports competitions.[135] In her experience, the teachers who "were older...they were patient but yet firm in their teaching," and equal with all:

> I don't think there was a difference because I was black and you were white. That attention wasn't given [by teachers] or that it was more demanding here and not there. I didn't see that.[136]

According to Dotty, if one didn't succeed in a subject matter, whether one was white or black, it wasn't because of the teacher, but because one hadn't put in the effort required of a student:

> If you didn't get it, it was because you didn't want to learn or you didn't want to take the time. So in that respect, I can say that it [academic standard] was demanding.[137]

In Dotty's memory, teachers were good disciplinarians who "would come out and really talk to the kids and break up fights"; they were hands-on authority figures who took care of their charges. Mr. Lancaster, who retired in 1964 after presiding over Miller High for fifteen years, was remembered fondly by Dotty, as he had been by all alumni, white and black.

While Dotty's memories tell of her feeling more a part of her high school than Doris or Annie ever did or could, and of her forging long-lasting friendships with black and white peers and participating in extracurricular activities, she still only spent two years at Miller High before leaving at the end of her tenth grade:

> My girlfriend got pregnant, I got pregnant, and we just ended up finishing, I finished my tenth grade. I ended up going to finish in the city and so did she; we went together and we finished our last year and graduated... It was more for embarrassment. If the times were like today, and people accepted more, I mean, it's nothing now. Then, you were stereotyped. 'Oh, she's having a baby, she's not married,' you know what I mean... Going back to school with friends that you were close to, you know, you don't want to do that.[138]

The tender subject of an adolescent's pregnancy and Dotty's genuine vulnerability and courage in sharing her story settled my dilemma of further probing into the nature of her embarrassment vis-à-vis friends. Dotty's rewarding beginning at Miller High leaves one imagining her high school career as a successful one. By the late 1960s, as yearbook data and informal interviews suggested, and as the career of Annie Cole's sister attests to, young black women attending Miller High would continue to succeed academically. Annie's sister, who graduated from Miller High in the early 1970s, went on to become a nurse.[139] By the end of the 1960s, life at Miller High for middle-class black young women had become more comfortable.

The lives of young middle-class black men who were stars on the playing fields in sports competitions continued relatively unchanged throughout the 1960s. They continued to hold a privileged status among peers as athletes, and at the very end of the 1960s, when football was introduced, as jocks, and they were most likely by virtue of this status to develop friendships with their white male peers.[140] Yearbooks reveal very few black adolescent males enrolled in the academic track throughout the 1960s. Those who were, however, were also jocks. Referred to by narrators who graduated in the 1950s as "athletes," jocks would continue to hold, throughout the second half of the twentieth century, a privileged position in the recollections and imaginations of Miller High alumni. Jock life was a place where black and white men collaborated.

For those middle-class young black men who neither attended the academic track nor were part of the jocks' crowd, life got a little easier. Nat Right, a 1963 graduate, remembered a Miller High where "after the first couple of years [immediately following desegregation], things started to mellow out, [and students] got down to being students."[141] Nat's traumatic experience when a white boy hit him in the mouth and he missed a full year of school while undergoing gum surgery did not stop him from pursuing his high school education. Still, while Nat thought eventually things calmed down between whites and blacks (especially following the Halloween dance incident), and while he remembered Principal Lancaster as "probably one of the fairest ones there,"[142] his overall experience continued to be one of discrimination.

> There was some discrimination... [teachers] ignored you if you raised your hand... we were not supposed to have knowledge. It was perceived back then that blacks were inferior... There was always somebody in your class who couldn't stand you because of your color... You had them in every class.

As Nat recalled, Miller High was still a fairly segregated place:

> Blacks tended to hang out with blacks, and whites tended to hang out with whites, and girls tended to hang out together, and boys tended to hang out together.[143]

By 1967, when African American alumnus Burt Sadden graduated, the transition into high school for young black men was much easier. Burt recalled that "it was pretty easy"[144] going from middle school, where he had already forged many friendships, to high school, where he continued those friendships. By the time Burt attended Miller High, and as Nat's testimonies suggested, things had changed, and blacks and whites had gotten used to working side by side. By the very end of the 1960s, some black youth perceived Miller High faculty as being fair. Burt remembered: "Back then it seemed to me that everybody was treated fair and equal."

Still, although Burt did not "recall too many racial type fights in school, just kids disagreeing," and although in his remembered experience all students were "pretty much treated the same" by all teachers, black and white students mostly continued to stay apart.

> Back then the blacks stayed on one side, the whites stayed on the other side. The blacks stayed to themselves, and the whites stayed to themselves. Everyone once in a while, when we had to, we would intermingle. Other than that it was the blacks stayed to themselves and the whites stayed to themselves.[145]

Unlike Dotty, who remembered a more pervasive intermingling of students across racial boundaries, Nat and Burt recalled a continued separation between white and black students, a pattern that indicated perhaps a greater degree of racial border crossing between black and white women in general tracks than between black and white males in the same tracks. Annie's and Doris's recollections of the late 1950s, immediately following desegregation, hinted at timid communication across racial divides between white and black female students; Dotty's recollections described friendships and bonds established between black and white girls in the general track by the mid-1960s. One constant remained, however: For the general-track black and white students, both males and females, Miller High was not a place you visited after class hours.[146]

Thus the upper academic track, as earlier suggested and as revealed again in Burt's recollection, continued to be populated by those students with means, or at least with the familial support to engage in all aspects of school life, and the lower general track continued to be populated by economically disadvantaged students who could not fully engage in the life of the school.

By the time Burt Sadden graduated, a decade after the Halloween dance incident, dances were now safe for black students to attend. But by then, however, as Burt remembered, blacks chose not to go to dances.

> We could have [gone to dances] if we chose to. We didn't feel comfortable with the music that was being played at the time. It was a music thing.[147]

Thus, although African American alumni Nat and Burt, graduates in the 1960s, recalled a continued separation between white and black students, they did not explain the separation in the same ways that graduates of the late 1950s and the very early years of desegregation explained it. For 1950s graduates Annie and Doris, it was a "black and white" thing. Animosities between black and white students were strictly defined by skin color. For Dotty, Nat, and Burt, graduates of the late 1960s, the "black and white thing" acquired a more complex cultural dimension. Beyond the question of the skin color, students hung out with students with whom they felt most at home, and in the general track, black students felt most at home, culturally, with other black students.[148]

For Nat, upper middle-class African American alumnus, and for Burt, working-class African American alumnus, both of whom attended the general track, high school was something you had to do and couldn't wait to get out of, just as it had been for white alumnus Robert Heart who was of lower socioeconomic background. Unlike Robert Heart, however, Nat and Burt were not out to "beat the system" and prove they could stick it out. For Nat and Burt, finishing high school was a question of honoring their parents and fulfilling familial obligations. Burt explained:

> I did attend graduation, of course; it meant to me that I achieved what I originally set out to achieve as far as my family was concerned. This was one of the things my mother and father emphasized to me: you had to be at least a high school graduate to be partially successful in this world. Back then that was pretty much true. We couldn't go to college. It was more important to try and get a decent job and survive.[149]

Black males in the lower general track endured school, and just like their white male counterparts in the general track, they couldn't wait to get out. Nat recalled: "I did as little as possible. I wanted to get out and keep on going".[150]

Students of the Divided Generation of the 1950s and 1960s, as alumni recollections revealed, identified their places vis-à-vis one another and school authorities primarily along the socially inherited binary constructions of black *or* white, female *or* male, middle class *or* poor. Constructions of the middle-class category broadly included lower-/working-class status, by contrast to the very poor on one extreme and the very rich on the other, as we shall see in the next chapter. Recollections across categories of social identity also underscored school authorities' leniency toward girls in general, and the influences of the combined force of tracking and economic realities of college attendance in dividing students within a matrix of intersecting categories of gender, race, class, upper academic, commercial, and general-track identities and labels.

Thus, in the 1950s and 1960s, white middle-class females and those who could play the part felt most at home at Miller High, where they enjoyed a myriad of opportunities for self-expression. Poor white adolescent girls, on the other hand, shunned by peers and teachers alike, often dropped out or became invisible. Middle-class black female students, who felt least comfortable within what was to them a hostile and alien institution in the early days following desegregation, transformed their circumstances from marginalized students attending at best the commercial tracks, to college-bound integrated students attending academic tracks by the very end of the 1960s. While the very few middle-class black male adolescents who attended the academic track and excelled in sports felt empowered by their education at Miller High, young white males on the other hand, whether in the academic or general tracks, whether middle class or poor, and many black males in the general track, experienced high school as a place where one "did time," as an irrelevant and economically useless institution, although, as we have seen, for different reasons across categories of race and class.

However, while structural forces largely circumscribed the relational nature of Miller High students' experiences with each other and school authorities in the 1950s and 1960s, by limiting meeting places between richer and poorer, between whites and blacks, and by organizing poorer white and middle-class black females in commercial and general tracks, and poorer white and middle-class black males in the general track, Miller High students, through the lenses of their situated positions, and within the setting of a rural town in the 1950s and 1960s, gingerly negotiated those positions. Under the range of unrelenting hierarchies, some students found ways to cross boundaries of class, race, and gender, as I explore in the next chapter, where I also explore the meanings attributed by students to the social categories by which they identified their place and that of others at Miller High.

Chapter 2

Cautiously Negotiating Social Divides: A Conservative Student Body (1950–1969)

Within divided and dividing institutional spaces, some male and female students in the 1950s and 1960s practiced crossing class and race borders in their imaginations when they could not do so in reality. Overall, however, students' opportunities for developing friendships across gender, race, and class divides were restricted and shaped by institutional norms. Some students willingly embraced those norms; other students, by dropping out or doing time, perpetuated and reinforced those norms, making the Divided Generation the most conservative of the three generations.

Peer Relationships

Intersections of Class, Race, and Tracks

Class was a category of identity to which students of the Divided Generation were particularly sensitive, whether they were black or white. Overall, adolescents who came from wealth did not attend Miller High, and Miller High students were very much aware of that. The very wealthy youth attended the many well-known private schools in the surrounding areas. African American young women and men might know about them through their mothers, whose domestic work brought them in contact with the area's wealthy families. Some young white girls of lesser means might know them through short-lived courtships they imagined might change their family fortunes.[1] These wealthy students occupied virtual spaces in the minds of many who attended Miller High. Their absence at school was recognized only in contrast to their presence in the neighborhoods. Graduates across the 1950s and 1960s recalled the contrast:

> We knew that kids who went to private schools were different. And I can remember when we worked on the yearbook, we would try to go out and collect money... we were like these poor little village children knocking on the castle door... This was just where the rich people lived, and the rich people were different from us.[2]
>
> The richer kids, or the kids whose parents had money, wouldn't necessarily send them to public schools; they sent them to [narrator names all the private schools in the area].[3]

It was also their absence at Miller High that made the school middle to lower class.

> Everybody was middle or lower class. I mean, we didn't have any upper class. They all went to private schools. If they had any money, they went to private schools.[4]

Thus, attending the public high school of Miller Town meant attending a school less likely to be frequented by the very rich. Within the middle- to lower-class continuum of Miller High, however, students arranged themselves and others as poor or upper class. Within high school borders, a status comparison quickly replaced the area's virtually absent upper class with a high school "upper class" defined by Judy Law, Robert Heart, and those who considered themselves "poor" as those who were college bound, drove their fathers' cars, and spent their after-school hours involved in extracurricular activities. Thus Alice Web, who felt like the "village child" when knocking on rich families' doors for yearbook donations, constituted in the minds of students like Judy Law and Robert Heart the "upper class" of Miller High—those who could afford to participate in after-school activities, and whose fathers could afford to pay college tuition.

Still, while the area's rich lived behind what seemed like "castle doors," within Miller High, some students practiced crossing class boundaries. Young white women of lesser means who could look the part forged friendships with more privileged white women within clubs held during school hours. Thus Judy Law and Alice Web became friends through yearbook club collaborations. Black young men in the upper academic track forged friendships with white males in classrooms and on sports fields after school, as the stories of Norman Good, Willie James, and Jimmy Cole revealed. Poor white farm boys could cross class lines briefly during their senior year, when they might date a junior "upper-class" girl.[5] Sometimes, however, poor white boys just couldn't wait. Now and then, they'd force themselves into the "upper class."

> Us lower peons, we just kind of hung out with ourselves...they [upper class], because they were going from one academic class to the next academic class...would walk down the corridor and they talked to themselves...and some of us would maybe be invited to their parties. Not very often...well, if the upper class didn't invite, we would just go anyhow. Because we would find out. We would go anyhow and show up.[6]

Thus, for white young women, the requisite for class-crossing was "looking the part." For black young men, the requisite was excelling in sports and academically within the academic track.

For the poor farm boy, the requisite was grade-level status, and, sometimes, bully behavior. The very poor whites, those who lived in shacks, were not reported to cross class lines.

Middle- and lower-middle-class black young women were, until the late 1960s, mostly funneled into the commercial tracks along with white young women of lesser means, and did not participate in extracurricular clubs such as the yearbook and school newspaper—clubs overwhelmingly populated at the time by white young women of means. They might have found opportunities to interact with the more privileged white young women only during physical education, held during school hours, since they were also less likely to attend after-school activities. Testimonies suggested that the black young women of Miller High in the 1950s and 1960s, by virtue of being assigned to lower academic tracks, mostly interacted with white young women of lesser means.

Class distinctions, as already suggested, were inextricably linked to racial differences. Although Nat was considered "aristocracy" among his black peers, he continued to be considered "poor" among his white peers in the general and vocational tracks. Testimonies suggest that at Miller High in the 1950s and 1960s, regardless of evidence to the contrary, if a youth's skin was black, then that youth was poor in the minds of the poorer whites. Speaking of his economic circumstances, Nat Right recalled:

> Blacks were considered poor among whites. As far as we were concerned, we were the aristocratic blacks...Blacks thought we were so rich.
> *Your family was considered to be wealthy, but you came to school and the white kids considered you poor?*
> Right.[8]

The two different testimonies by African American alumnus Norman Good, who attended the upper academic track, and African American alumnus Nat Right, who attended the lower track, suggest greater prevalence in the lower tracks of intolerance among whites toward blacks. Norman recalled experiencing acceptance by richer whites in the academic track, while Nat recalled being rejected by white peers in the general track, even when his place of residence proved his economic standing.[9]

Norman's and Nat's contrasting testimonies bring to light poorer white students' denial of black students' possible economic success. The high status that the academic track held in the school to begin with, coupled with its demographic composition (overwhelmingly white females and males of means with varying levels of academic credentials, but only one or two black students with impeccable academic credentials), might have eliminated the pressures of race competition in the wealthier upper academic track. In the lower tracks, however, as testimonies suggest, perceptions of class differences were inextricably linked to the question of race. During the early years of desegregation at Miller High, as earlier shown, racial encounters between young black and white men in the lower academic tracks were tense and even violent.

It is important to underscore that a young black man entered the rarified air of the upper academic track only through persistent parental insistence, relentless self-discipline, and superior performance in academics and sports. While tracking funneled white students with means regardless of their academic merits into the upper academic track, a black youth with means would be systematically assigned to the general track. Thus a young black youth's class status was trumped by race. In the eyes of white administrators as well as white peers, particularly male peers of lesser means in the general or vocational tracks, a black youth was black first and only.

Male peers in the general track did not accept Nat Right's middle-class status. The general and vocational tracks were where some of the angriest and most disenfranchised young white men of the rural community faced the company of young black men, most of whom, while disenfranchised within the broader white community of their time, were by contrast fully integrated into their black community and steeped in family values that emphasized respect for parental authority and familial obligations. Resistance to Nat Right's middle-class status by white peers in the lower tracks might suggest that students pejoratively called "hillbillies" by middle-class white peers felt doubly threatened by black youth who held many of the same values—and the same Christian affiliations—as middle-class white peers.[10] Denying black young men middle-class status would have negated the possibility of their social superiority in a world where the poor white youth were considered the least for having the least.

Intersections of Gender and Race

Whites and blacks more easily crossed racial boundaries within same-gender relations. However, crossing racial and gender lines simultaneously was problematic. Although encounters between young white women and young black men began emerging in the 1960s, as Dotty's recollection suggested, interracial coupling at Miller High was taboo throughout the 1950s and 1960s. Black and white students who might have been attracted to each other kept their desires secret. Doris Right, a black female alumna, recalled a white male classmate confiding to her years later: "I used to watch you in class. I really liked you, but you know, we couldn't say anything." Doris also confided her own repressed interests: "You know, you would look at a person [white male] and say, 'Oh, gee, he's cute,' and that would be it. You wouldn't even entertain the idea, you know, of any type of relationship." By the mid-1960s, one rare student couple ventured crossing racial and gender borders at Miller High. Dotty recalled that the young man was black and the young woman was white. Still, the pair's dating was more private than public, and the couple stayed away from school socials.

It is perhaps no wonder, given the social taboos surrounding interracial couplings, that racial encounters would be most violent at high school social events that involved dates. African American alumni recollections of the Halloween dance vividly depicted the tensions involved.[11] When it came to the dance floor, social status quo of the time prevailed. White racist students, buttressed by their white teachers' tacit support and significantly outnumbering black students,[12] claimed full control over the dance floor.[13]

Social status quo further prevailed in the power differentials manifested in dating choices. White males of means exercised the greatest choice in terms of whom they could date among white young women, since they could date white women of lesser means, as Judy Law's testimony showed. Next to them, white young women of means exercised the most choice by actually rejecting poorer boys, and accepting their advances only if the young men were seniors and they themselves were of lower grade level, as Robert Heart's recollections suggested. Black males and females exercised the least choice when it came to dating within their high school community because there were too few blacks to form a big enough social network for dating.

Beyond dating and the dance floor, power differentials in gender relations within everyday high school life were different for white and black students. White girls, graduates' stories have suggested, patronized white boys, whether by doing their homework or taking charge of all social activities related to the school, and the boys allowed them as much. In relating with one another at Miller High, white males and females reflected behavior more akin to the private realm, where, during the 1950s, as historians have noted, "women could be a solution to men's dilemma...[and] provide men a haven in a heartless world,"[14] suggesting a kind of domestic takeover by white middle-class female students at Miller High. White boys let them have the run of things, not unlike white men of the time who let their wives have the run of all things domestic, including volunteer work.

As historian Linda Eisenman suggests, "The 1950s encouraged women's activism through...civic minded organizations such as the League of Women Voters...Young Women's Christian Association...Women could participate without committing themselves to regular, paid employment."[15] While Miller High's white boys lived on the edge of classroom participation, working before and after school, sprinting out of high school windows to join firefighters at a moment's notice, or taking leave to go hunting, white female students were stable residents. When they did leave school property, it was to go on errands to shop with money already provided them by their elderly teachers. A short leap of logic explains the seeming disjuncture between young women holding leadership positions in the Future Business Leaders of America and on the editorial staff of the school newspaper without corresponding leadership positions in the world outside high school boundaries. As long as the work remained unpaid, white young women were encouraged to develop their leadership roles.

This also helps explain why white boys of lesser means who were not college bound placed little value on attending classes for which one received no pay. For them, the "real" work occurred outside of high school in the "real" world. So much so that some of them risked, in a gesture of ultimate disempowerment of school authority, punching out the demanding teacher and forsaking school altogether for the possibility of fortunes made in the "real" world. Because college, which held promises of money and ascendancy into upper-middle class, required money to attend, being a student did not pay for poorer white boys. As for "richer" white boys, if not college, then jobs in their fathers' businesses often awaited them, suggesting a continuation into the 1950s and 1960s of a pattern of nonchalant attendance among white male students, which historians of education have identified with the beginning of the high school in the early nineteenth century. Regardless of whether they graduated from high school, these young men were guaranteed jobs.[16] In general, college-bound white girls and boys prepared for "middle-class families [that] came to dominate the era...a white-collar husband, supportive wife, and several children residing in a comfortable home."[17] Judy Law explained how "you were meant to grow up, marry the boy next door, go to the local church, and stay in the family."

By contrast to the lives of middle-class white young women, middle- and lower-middle-class African American young women headed straight home after school. During the early years of desegregation, they might have sought the help of their more popular male counterparts to stop some white boys' misbehavior toward them. However, white boys' more offensive behavior toward black girls were furtive, and the culprits went unrecognized as they anonymously and perversely "felt behinds" in the hallways. For middle-class young black women of the early days of desegregation, school was not a safe place. Whether sexually offensive[18] or less sexual and more annoying, as in DiPaglia's teasing of Annie, some white boys' behavior toward some black girls at Miller were experienced by black girls as aggressive and sexual. But it wasn't just white boys' aggressions they dealt with. Some black girls fought with black boys and black girls because they looked different. Doris Right remembered:

> Some of the black kids didn't like us... We were different, not that we knew it, we didn't know that we were any different that any other black kids. My sister and I have the green eyes—nobody during that time had green eyes—and they would call us "gray-eyed." Whatever, they weren't even gray, they were green, and they'd fight us because the color of our eyes... And one guy, he was a big guy... he would want to fight me every day.[19]

Doris's description of her relationship with other blacks echoed historian Franklin Frazier's descriptions of black students in the earlier part of the twentieth century, when "discriminatory practices were visited by light-skinned students on darker ones."[20] However, in the case of Doris, something else might have been at play that explained the discrimination she experienced by other black youth, since Doris's black skin is darker rather than lighter. Doris and her sister, although light-eyed, were also the daughters of parents who owned land in the more prosperous part of the town. Middle-class black female students in the early years of desegregation at Miller High experienced their bodies as targets for white boys' furtive stabs at sexual misconduct or for other blacks' assaults. That their persons, their emotional and rational selves, were submerged under the combined weight of their gender and skin color in the early days of desegregation was made even more apparent in contrast to the lack of reported aggression toward white women. Immediately following desegregation, racial inequities were compounded by gender inequities for Miller High's young black women. By the time Dotty attended high school in the mid-1960s, as earlier discussed, life at Miller High for young black women felt more comfortable.

Intersections of Race and Extracurricular Spaces

Against resistance by many whites who never ate with blacks at lunchtime, white and black students would occasionally sit together at lunch. Either the black student was a star athlete who sat with white star athletes, or pockets of white students were neighborhood friends of some black students and had lunch with their childhood playmates. Speaking of her white friends, Annie Cole recalled:

> You would have a few that would intermingle with you [in the cafeteria]. Like the kids that we played with in the neighborhood would come over and sit with us. Like Ike and Dewey and Charlie, and all of them... Some of it was nice. We weren't all bad.[21]

Knowing your neighbor made sitting with her at lunch a natural extension of your life in the community. Although blacks were still not allowed in restaurants reserved for whites during the early years of school desegregation, some Miller High black and white students were eating together at the same table in the public setting of the high school cafeteria. There they could practice new social mores within the protection of the law. However, young blacks and whites overwhelmingly segregated during lunchtime, as underscored by African American alumnus Nat, who attended the general track, and white alumnus Bud, who was college bound. Thus the cafeteria, the one place within the high school that housed the greatest numbers of male and female students of different racial and economic backgrounds in one place at one time, never fully became a place of integration, even as some students ventured to cross identity boundaries. On the contrary, it would become, as we shall see in later years, the barometer for levels, not of integration, but of segregation (except for the Border-Crossing Generation), and a place more akin to a prison mess hall than a young people's lunch area.

It was on the sports field after school that black and white males practiced becoming friends. Black and white females experienced friendly relationships on sports fields in physical education during school hours.[22] The one space within Miller High where students most consistently crossed racial borders (but never gender borders) throughout the 1950s and 1960s was in team sports. However, access to team sports after school required student availability, which poorer students didn't have, and which middle-class black young women did not have during the early days of desegregation, because their parents requested their presence at home immediately after school for reasons of safety.

Implications of School-Shaped Associations for Cross-Racial and Cross-Class Encounters in the Workforce

While relationships across gender proved more empowering for white middle-class young women and less empowering for black young women, relationships within gender proved to be more conciliatory among white and black women in the commercial and general tracks than among white and black males in the general tracks. Non-college-bound young women, black and white, were more likely to be enrolled in the commercial track. In view of the fact that preferential treatment was given to girls, as reported by white and black male and female graduates, it is not surprising that in general young women might have had an easier time associating with each other across racial boundaries within lower academic tracks.

However, while the tracking system funneled poorer young white men and black young men regardless of economic background into lower curricular tracks that prepared them for menial labor, it funneled young white women of lesser means and black young women across economic backgrounds into lower curricular tracks that prepared them for secretarial white-collar jobs. This gender division across race in lower tracks opens up for investigation the implications of gendered differences in tracking in the 1950s and 1960s (females {black and white} = commercial; males {black and white} = vocational) for relationships continued into the workforce—privileging perhaps white and black women's opportunities for crossing class lines over white and black men's opportunities to do the same. Secretarial work generally brings one in contact with people of different economic backgrounds, while menial labor work mostly keeps one isolated from daily contact with people of upper-class backgrounds. Given the flourishing small black-owned businesses of Miller Town in the 1950s and 1960s, there would have been opportunities for young black women to find employment as secretaries and to encounter people across both class and racial lines. Further research is needed to investigate whether all women have greater opportunities than men to establish relationships across class and race lines in the workplace at large.

Youth and Authority

Constructing Teacher-Student Relationships across Social Categories

Overwhelmingly, this was a teacher-centered generation. Except for college-bound white alumnus Bud Land, who graduated in 1968, all alumni interviewed recalled particular teachers who helped them and who eased the difficulties of high school life: the gym teacher for lower-middle-class African American alumna Annie Cole; the shop teacher for poor white alumnus Robert Heart; the older female teachers for poor white alumna Judy Law; the coach and journalism teachers for college-bound white middle-class Linda Moss; and so on. In some graduates' reports, favorite teachers taught subjects that students loved and excelled in;[23] in other reports, favorite teachers were just or inspiring, even when students neither loved nor excelled in the subject matter. Sometimes the teachers were close to the students in social status, as in the case of the shop teacher and Robert Heart; or close emotionally and in world views, as in the case of Annie Cole and the gym teacher. Annie Cole loved her gym teacher not only because she appreciated Annie's athletic skills, but also because she treated blacks and whites equally and was someone in whom Annie could confide. Judy Law felt sustained by her elderly white teachers because in their company she was appreciated for the middle-class things Judy longed for and strived to live up to, things she was denied in her isolated and poor country life at home: looking pretty, writing poetry, winning awards. Linda Moss loved the teachers who believed in her academic abilities, and so on. These hero-teachers often filled voids in students' lives, helped them to see worlds beyond their own, and to gain freedom in imagination if not in reality. In the 1950s and 1960s, in a mostly rural small town, with a very small high school population and an economic and social climate conducive to class mobility, the individual teacher could become the agent of change that interposed motivation, inspiration, vision, and fairness between a student's social background and the comprehensive high school's often-documented bureaucratic tendencies for social reproduction.[24]

However, good relationships with teachers did not necessarily translate into a good education for all students. The mosaic of recollections from white graduates of the 1950s and 1960s sketched a high school education that either failed to prepare them for life after high school or was irrelevant to life outside of high school. College-bound graduates were surprised and shocked at the level and quality of work required at the college level. They felt completely unprepared for higher education, and graduated from Miller High with few study skills and a weak academic foundation.[25] By contrast, African American alumnus Norman Good, who attended the academic track, remembered receiving a solid education: "I feel as though I got an excellent education at Miller." Recollections, as earlier discussed, also revealed that among the college-bound white students, excelling in academics was not as crucial as it might have been for a black student who would have had to invest every ounce of effort to prove that he was worthy of the academic track. Accordingly, his efforts would have paid off in academic achievement and scholarly preparedness. Thus the academic track for white students was less about academic aptitudes and primarily about economics, particularly for males.

To both college-bound and non-college-bound white males, academics seemed irrelevant to "real life" as they knew it in Miller Town. White males of means would, as documented by historians for generations of high school students before them, inherit their fathers' businesses, or "become clerks, tellers, salesmen, or agents, and were thus able to maintain their socio-economic status without the effort of long term study."[26] However, although by the 1950s young white men could drop out without much economic repercussion, they could no longer drop out voluntarily without negative social repercussions.[27] Thus, many "did time." Their "real lives" were defined by their fathers' businesses. White males of lesser means also did time as they continually juggled jobs on the side. They too perceived their academic education to be irrelevant and disconnected from the realities of their lives. These white male students, richer and poorer, dissatisfied with the "system," turned out to be perhaps Miller High's most conservative force. Not only did they "suffer" the academic system, sometimes out of loyalty to their parents, and relinquish the school's social life to their female counterparts, but also inadvertently supported, by their complaints of the irrelevance of an academic curriculum, the public high school status quo since the triumph of the comprehensive high school.[28]

During the 1950s and 1960s at Miller High, high school education would prove most valuable to non-college-bound white and black women and to the very few black males who attended the academic tracks. For these students, the high school diploma meant possibilities for social mobility and expanded terrains of freedom. These students also proved to be a conservative force in the perpetuation of the comprehensive high school status quo, because they invested their energies in using its education within its prescribed modus operandi to better their lives.

A Conservative Student Body across Categories of Social Difference

On the whole, students at Miller High acted much more conservatively than sociological case studies of 1950s and early 1960s high school life suggested. Coleman, the education ethnographer most notable for identifying high school youth of the 1950s as focused solely on popularity status among peers, painted them as primarily engaged in their own adolescent societies, where "looking good" determined hierarchies of popularity. He underscored how little influence adults held in the lives of high school adolescents.[29] This work, on the contrary, reveals young white females' negotiations of appearance as a means to maintain closeness with teachers and school authorities who more readily accepted a certain look in their charges. As alumni reports revealed, young white women's behavior at Miller High was not as much about popularity among peers as about maintaining good relationships with teachers and practicing being pretty and successful. Acceptance by authority figures whom they considered role models was of utmost importance to them. These white young women were a very conservative force, as Coleman's study suggested, but at Miller High they strove to please authority figures with whom they shared almost familial relations. Their focus was on ensuring their particular *student* status, rather than *peer* status.

Coleman's assertion that "a working class boy or girl will be most left out in an upper middle class school"[30] does hold true in this work, evidenced in hierarchies of dating patterns and Robert Heart's acute awareness of class distinctions. However, at Miller High, as alumni testimonies suggested, students—female and male, black and white—befriended or shunned each other within the predetermined tracking structures, suggesting that the school's organization of students within predetermined groupings influenced to a much greater degree their associations with each other than their own volitions. In fact, as recollections of white and black alumni revealed in this work, many students wished to associate with each other across dimensions of class, gender, and race, but, as Alice Web expressed, "We were to a certain extent in our slots."

On the whole, students of the Divided Generation participated in perpetuating the high school life of the early twentieth century, where "commitments to competition, conformity and individual merit" prevailed,[31] by the traditions they embraced from previous generations, and by what they didn't contest. The "doers," as Linda Moss called them, made sure to muster school spirit, to organize Miller High's social life, and to please their elder role models. Black males in the academic track relished the opportunity to compete with their white counterparts and to prove themselves. Poorer students and those who experienced school as an unfriendly if not outright hostile place dropped out, leaving the run of things to the "doers." Students for whom school was irrelevant or hostile, but who stayed, "did their time." Although the story of Miller High students' lives during the 1950s and 1960s is one of many intertwined stories, on the whole the school's transmitted hierarchical divisions heavily organized their lives.

While a fair and benevolent principal helped the small student population navigate the highly charged period of desegregation with less rather than more damage, and while a very few teachers helped build hope and self-esteem in students across social categories, students lived lives mostly segregated along gender, race, and class divides. Entering the world of Miller High in the 1950s and 1960s meant stepping into the habits and traditions transmitted and upheld by white middle-class female students and teachers. It was a high school ill-equipped to help students relate with each other across categories of identity, even though it offered greater possibilities for democratic engagements among students across racial, class, and gender divides than the rural town it served. It was where students by and large remained close to their own across categories of identity.

School structure, rather than youth culture, overwhelmingly determined how Miller High students understood their place within school and how they imagined their place outside of school in the 1950s and 1960s.[32] Overwhelmingly, alumni recollections revealed that tracking divided students by class, gender, and race. Tracking had divided students by class and gender before desegregation in 1956, and then by race following desegregation and throughout the 1960s.

Carnoy and Levin's suggestion that high school has consistently provided a "social experience that is markedly more egalitarian" than anything the workforce might provide does not hold in the case of many students of Miller High in the 1950s and 1960s. This was not the case for white male students of privilege, whose work opportunities were guaranteed regardless of success in school; nor was it the case for poorer white males or middle-class African American young men who attended the lower tracks. For poorer youth, as recounted by those interviewed, school was a place of shame and took time away from earning sorely needed money. For the black male youth attending the lower tracks, Miller High became a place to suffer the full-blown force of racism. While a lone interracial couple would emerge at the end of the 1960s, and a very few students crossed race divides in an otherwise overwhelmingly segregated cafeteria, students by and large segregated by gender, class, and race.

Carnoy and Levin's observation does not however apply to white middle-class girls who practiced roles denied to them in the workforce. Overall, Miller High in the 1950s and 1960s most consistently and effectively privileged white middle-class adolescent women more than any other student constituency, since they would prove best suited to play roles agreeable to white authority figures, and to reproduce within and for Miller High the dominance of a white, conservative, middle-class lifestyle.

But by the end of the 1960s in Miller Town, with the introduction of football and marijuana, came the end of an era of formal manners and clear hierarchical relationships. Beginning with the 1970s, *in loco parentis* would no longer guide teacher-student relationships. As one alumna, a 1969 graduate, remembered: "We were the last class that wore skirts. The class after us, the class of 1970, really changed a lot of things."[33]

Part II

The Border-Crossing Generation (1970–1985)

Chapter 3

Memories of Interracial Peer-Group Affiliations: Integration Years (1970–1985)

In the 1970s and early 1980s, Miller High students began their school year as generations before them had: with the ringing of the bell by the faculty member with the most tenure. However, while they also inherited the *Key* yearbook, the *Miller Chronicle*, and the proms, the students looked and behaved in ways that often shocked the aging and retiring teachers. Young women could now wear pants, now and then some students might "streak—take all their clothes off and run through crowds,"[1] some young men wore their hair long, and some African American young men and women wore theirs in Afro style.

By 1972, sixteen graduating classes had attended desegregated schools. Moreover, while Miller High elementary feeder schools were mostly populated by white students and only handfuls of black students, some white children grew up as minority students in the one predominantly black elementary school in the mostly black neighborhood. As one white alumnus recalled:

> My friends in first grade were black. I went to the black school. After everything was integrated...they put the overflow [white] kids in the annex on Brand Avenue...They [the black kids] were my first friends. When I see those girls, Maggie and Loretta—I saw Loretta two weeks ago—it was just the greatest thing in the world![2]

While still known for its predominantly rural flavor, throughout the 1970s and early 1980s Miller Town was developing into a suburban settlement with a variety of housing types, from single-family homes along sidewalked roads perpendicular to Main Street to ranchers in gated communities for business executives. It was also in the early 1970s that black families' houses on Brand Avenue finally received indoor plumbing. The "shacks" in which the very poor lived, as reported by 1950s graduates, no longer existed in Miller Town by the 1970s; the poorer whites lived in rented apartments, often above shops or taverns along Main Street. Miller Town's population became increasingly white and suburban during this period, with the intensification of white flight from city limits, and "the local people, both black and white,"[3] people whose families had lived in Miller Town for several generations, began uniting against newcomers. This bonding of "old timers" would continue to intensify and reach, by the 1990s, a defiant tone, marking insiders from outsiders.

Within the walls of Miller High, the Future Farmers of America club was no longer offered; home economics was renamed bachelor living and offered equally to boys and girls; Future Nurses of America was renamed Health Careers Club; and sex education and drug prevention classes were added at the end of the 1970s. It is also during this period that younger teachers were hired to replace the retiring "old timers." This change created a division between young and old teachers, as young teachers demanded more academically but less behaviorally, and older teachers demanded more behaviorally and much less academically. Furthermore, the infusion of younger teachers, as alumni testimonies suggested, participated in destabilizing power relations between teachers and students: older teachers were less respected, and younger teachers sometimes crossed borders of familiarity to the point of having affairs with students.

During this period, nestled between the early years of desegregation and the immigrant wave of the late 1980s and the 1990s that flooded Miller Town with foreigners and city escapees, Miller High's racial composition fluctuated. With the intensified white flight in the early 1970s, the African American population decreased to two percent of the total Miller Town population,[4] compared to eight percent in 1950. By 1980, however, the relative population of black residents substantially increased to thirteen percent of the total population of Miller Town as white flight stabilized. Other things also changed—the educational backgrounds of Miller High attendees during that period were qualitatively different from those of the previous generation. These black and white students attended school together since the elementary level and doubled the size of Miller High graduating classes. By the 1970s, while many schools across the country were only beginning to desegregate, Miller High students had grown up in desegregated feeder schools. They knew each other not only as neighbors whose lives intersected on the streets and in the market place but also as classmates who had forged interracial relationships.

This generation further differentiated itself from the previous one in that alumni remembered one another less in terms of class and race and more in terms of peer groups consisting of jocks, preps, eggheads, potheads, motorheads, fireheads, musicians, and so on. This period also began with the introduction of football in 1969, which had been banned by the school throughout most of the twentieth century, as it was considered too dangerous for young men.[5]

The introduction of football, in turn, brought to prominence male football players, who, during the 1970s and early 1980s at Miller High, were particularly favored and remembered specifically as jocks: "The jocks were the group who played football."[6] The absence of football throughout the 1950s and 1960s most likely explains the rare use of the term *jock* by alumni who graduated before 1970.[7] All interviewed alumni from this in-between period that defines the Border-Crossing Generation recalled male jocks, black and white, as holding high-profile positions in the daily life of high school. To understand Miller High students' differently situated experiences during this period, one must first understand the high school world that black and white, male and female jocks participated in creating, for it is against this backdrop that the potheads contrasted, and the orbit around which the musicians, eggheads, and others revolved. As we shall see, it is also at the very periphery of all peer-generated groups that students who did not belong to a group continued to segregate by race and class.

Interracial Male Jocks

African American David Randle and white alumnus Josh White, both graduates of 1976, and Jim Garnes and Tim Whittle, white graduates of 1978 and 1981, remembered their high school careers not as students, but as jocks.[8] For these young men, the first day of school was an easy transition from middle school. Having trained together throughout the latter part of the summer, they were at home with one another before classes even started:

> We played football, so we knew each other. We got there in August. In July we'd do a weightlifting thing, start practice by the time school started...we were already acclimated. I knew what my classes are. You get your schedule ahead of time. It was very easy to start school.[9]

As testimonies revealed, a jock, whether white or black, richer or poorer, sat at the jocks' table in the cafeteria. When non-jock alumni reflected on their jock peers, they remembered them as a group unto themselves, those whom other males in particular perceived as having more fun. Musician Michael Hallner, graduate of 1976, recalled wistfully: "The jocks didn't hold back, they had fun...dated the best-looking girls in school." An indirect peek at Ronnie, a black jock graduate of 1974, through Josh's recollections, revealed a black jock fully aware of his hierarchical place, rendered by his jock status as well as his formidable physical strength.

> I remember one time I was walking through the halls in high school, and this black guy, *big* [alumnus's emphasis] black guy, played football, he was standing in the middle of the hallway, and he was intimidating everybody. He wouldn't let any one walk by. He was just playing around. I knew him, I knew he knew my dad [sports coach] pretty good, so I walked up to him and said, "Hi, Ronnie, how you doin?" He said, "I'm not touching you, your old man will kick my butt."
> *Were people intimidated by Ronnie standing there?*
> Yeah. He was big and all, but he was kind of a teddy bear too. Not that I would tell him that.[10]

Alumni who had been jocks admitted enjoying a highly privileged position in school:

> It was a huge social stature to be on a football team and being a good football player. I had one class I didn't pass anything, it was chemistry, I got a C all year and didn't do anything. I know that's why I got it, so I could keep playing football...Athletes back then did get away with things.
> *Did you take advantage of that?*
> Heck, yeah! We got to use the private bathrooms where the other kids couldn't. Coach let me have the keys.[11]

While David, Tim, Jim, and Josh shared memories of fun times as jocks, their recollections of non-jock peers differed, particularly for Tim, a white jock, and David, a black jock. Tim's recollections revealed a sense of righteous responsibility, and David's revealed a sense of systemic injustice and regret.

Tim, a 1981 graduate, remembered himself and his friends as protectors of people he considered weaker, among them the special education students and the girls. Tim's recollections suggested a self-appointed romanticized role not unlike a white knight—a righteous jock righting wrongs, fighting off the bad jocks:

> "You had a few kids that'd run their mouths always, assholes of the school. Sad to say, most of the time they were athletes, which was embarrassing for me. I got into a fight with a couple of them. Because they were teasing kids. I'd never do that. My friends and I never did that. You don't go fight the mentally challenged. That was taboo. Hands off. I got into a fight with some guy because he was teasing a girl, the way she walked. I said, "Leave her alone." He pushed me. I punched him.

These jocks, as Tim depicted, were also the protectors of blacks against white racists. Tim remembered:

> Me and Jack and Mat [came across] this guy [who] said, "We don't want them n— in here." [Me and Jack] beat the shit out of him, we beat him up in the parking lot, and Mat (the black friend) didn't even touch him.[12]

African American alumnus David Randle, who graduated five years before Tim, corroborated Tim's testimony, suggesting that Tim and his friends did not act in isolation. David recalled:

> [Racial slurs] were kind of taboo in that day. They would get on each other—I'm talking about white kids. They didn't even use the word. You know, it was a problem for them.[13]

In contrast to Tim's memories, however, David's recollections of other black non-jock peers were filled with regret. While Tim walked around "beating the shit out" of bad guys, rescuing those he considered weaker, and protecting the honor of his black friends, David spent much of his emotional energy explaining to his black peers why he played ball. David recalled:

> The people my age they were, I don't want to say militant, but more active into their culture. They would ask me, "David, why you playing ball for them, man?" I'd say, "'Cause *I want* [David's emphasis] to play ball." "They're just using you," they'd say.[14]

David's testimony revealed a frustration with the white and black faculty who did not do more to actively recruit capable black young men into sports; young men who, as he saw it, needed to be sought out rather than left on their own to join a system they understood to be working against their best interests. David's recollections emphasized the importance of one's involvement in an integrated team sport as a means not only of self-expression, but also as a means for overcoming racial divides.

> It doesn't seem right to say that some of my best friends are white, don't sound right, but I don't know a better way of saying it. See, sports is different. You do bond there. You don't lose that. You gotta trust one another. Race doesn't play a factor into that when you get into a football team.[15]

David's love of sports competed with his loyalty to black peers. However, it was also his privileged position as jock in general, and the bonds that it allowed him to develop with white peers, that created the opportunities for David to introduce his black peers into the world of middle- and upper-middle-class whites:

> They [the rich white kids] had nice parties...I got in because I played on the team...I used to go by myself to the parties, but by senior year I used to bring other black people.
> *Was that a good thing?*
> That was a good thing. That was a good thing...it brought everybody closer. Then you go back to school Monday morning, you just spent all Friday night, all Saturday night together, it would bring everybody closer together.[16]

By his senior year, David had begun using his unique position as a black jock to actively create opportunities for black peers to cross racial divides. Not unlike Tim, who somehow felt responsible for the well being of those unjustly bullied or discriminated against, David felt responsible for his less privileged black male peers. None of the jocks interviewed reported thinking much about life after high school, and all of them attended school willingly, even happily. However, how and why they made it through school differed for David from his white peers' experiences.

For David, going to school meant going against the current of truancy among his non-jock black peers. He explained:

> I loved high school. I went to school every day. Some of my friends, they'd cut school all the time. [After some time] they knew I was going to go to school. They wouldn't even say, "David, are you coming with us?"[17]

David's genuine enjoyment of high school might not have been idiosyncratic. Other graduates' recollections suggested a shared playfulness and ease among black jocks at Miller High during the 1970s. Resisting peers' truancies seemed easy for David. It was outweighed by his passion for football and acceptance by white peers.[18] David explained his non-jock black peers' alienation as the result of an unresponsive school system coupled with an upbringing that did not discourage using race as an excuse for not participating in school:

> I think those kids that the system beat them...for one reason or another the teachers kind of brushed off [the black kids]...A lot of black kids didn't participate in extracurricular activities. Nobody really approached us "do you want to join this or that?"
> I made it 'cause of my mother. "Don't come home tell me the teacher did this or did that"...the teacher was right no matter what happened. We were never allowed to use race as an excuse...If you want to do it, do it.[19]

David's willing attendance at high school, however, was solely and primarily linked to his playing football. His recollections of Miller High teachers were critical.

> For me it wasn't a very positive relationship with teachers. They kind of brushed us [the black kids] off a little bit, and we were just like there...I never met a guidance counselor in high school. I mean, they had my grades, they knew I wasn't stupid, but they never mentioned college or anything.[20]

Not only did he remember feeling frustrated with the mostly white faculty who ignored black students, but also with what he called the "wrong attitude" of the black male teacher who taught black history.

> He had a good old boy attitude...He even told us, "They're making me teach this class for obvious reasons."...But he didn't really get into black history. He could have been a lot stronger.[21]

David's recollections of Miller High faculty were far more critical than those of black graduates in the Divided Generation, suggesting a greater sense of rights to better services on the part of black graduates of the Border-Crossing Generation than previous black generations of Miller High students. David's insistence that faculty failed to reach out to black students resonates with the late-1950s poorer white males' demands for sensitivity to students' needs, a theme also echoed, as we shall see, by Tim Whittle, a working-class non-college-bound white jock. While David enjoyed a privileged status among peers and with the coaches, he continued to feel marginalized as a black student. His words, "I mean, they had my grades, they knew I wasn't stupid, but they never mentioned college or anything," echoed Norman Good's disappointment with not having been given options by guidance counselors to go to colleges other than the historically all-black colleges. The difference between David, a 1976 graduate, and Norman, a 1959 graduate, was that David expressed exasperation while Norman expressed sadness. Although frustrated with Miller High teachers, there was never any question in David's mind that he would not graduate. As many graduates across the two generations reported, honoring parents' wishes was a big part of graduating: "I never thought about not graduating... I think it was a big deal for the parents." Although David knew that he had to graduate, he never thought about what he would do once he did graduate. It was David's coach, Luke Dare, who suggested to David that he consider college and who helped him to apply:

> Luke Dare, my football coach, he just said, 'David, what are you going to do with yourself?' He said, "You going to college?" I said I hadn't thought about it. He said, "You need to think about it." And he set it up. He called the school and got the [football] recruiters to come over, and he took me over to the college and he introduced me to the football coach... I probably would have gone down one of those factories in Miller Town, but I went to college.[22]

David's status as jock and high-performing ball player opened doors into a world he hadn't previously considered. Although his life path turned out much better than he had envisioned, David lamented that many of his black peers whom he considered more deserving than himself did not participate in sports, where they could have shined and, as he remembered, could have enriched Miller High. David explained:

> There were good athletes from my neighborhood, but nobody really approached us about do you want to join..., athletes that were better than me... The school lost a lot by not getting more of the black kids involved.

David's recollection of his lesser athletic abilities but full extracurricular involvement compared to the superior abilities of his black peers on the one hand, but their lack of school participation on the other, underscored the precarious nature of high school success for black young men. As David understood it, it was a success that required, beyond mere attendance, an immersion in sports life both before and after school. Moreover, it required that the student actively seek participation in the absence of school authorities' invitations to do so. Thus the onus for success at Miller High lay squarely and solely on the shoulders of the black male student. Were black students determined to seek inclusion, they would not be refused, but neither would they be recruited. David recalled:

> Nobody approached us about... so what happened... a lot of kids, they didn't get involved in anything extra. They just wanted to get out of there.

2:15, time to go, we're out of here... It wasn't fun for them to go to school. I can tell you that. That's why they didn't go.

For David, a high school education became a passport out of factory life into higher education through a football scholarship that changed the course of his life, but for Tim, Jim, and Josh, it became a way to avoid hard choices and ultimately forfeit an academic education, choices that were facilitated by a laissez-faire older faculty and lax graduation requirements. Echoing the stories of white male graduates of the Divided Generation, white jocks of the Border-Crossing Generation continued white young men's nonchalant attitude vis-à-vis studies and reliance on their white female counterparts to do their daily homework. Tim recalled:

> I didn't have to work hard. Girls did my homework. I never read a book in high school. I'm embarrassed, kind of, I still don't like to read... To sit and read a book, I sooner chop my finger off. It is the most boring thing in the world for me. I never read a book. I never did homework. I didn't have to do it... So, far as I was concerned, my education was poor.[23]

Their status as jocks accentuated the ease with which they got through the system and with hardly any effort invested in learning. In large part it was the structure of the comprehensive high school that made it easy for them to buy free time as students by accumulating certain amounts of credit points.[24]

Miller High white male jocks of the Border-Crossing Generation, richer or poorer, knew they would find jobs after high school whether they attended college or not. Tim, who was the poorer jock and whose parents could not afford college, went to work for a well-established company in the area where he climbed the salary ladder. Jim Garnes went to work for his father and eventually took over the family's small business. High school education for the young white male jock of the 1970s and early 1980s, as it had been for white young males in general at Miller High at least since the 1950s,[25] was not about academic learning. Just as teachers in the 1950s looked the other way during hunting season, or when boys climbed out of windows to run to the fire station, in the 1970s, many of the older teachers' lax academic expectations conspired to lessen the importance and value of academic learning in the minds of their white male students.[26]

Jock Jeremy Garnes remembered:

> At high school level [the courses] were not useful and I could have done without high school... I remember taking a physics course, it was literally a joke... we would play the whole time... came test time, we never

studied. He'd give us a test and walk out the room. He had copies of the test out there. One or two people would go out there and tell us the answers. Back then the older teachers just wanted to...get done.[27]

Teachers who were older, according to testimonies, gave easy grades, but also were more likely to send their charges to the principal's office for misbehavior. Younger teachers challenged their students academically but were more lax with discipline.

Lax academic standards on the part of the older teachers contributed to Miller High's own version of what Labaree identified as credential inflation, and a reciprocal laissez-faire academic attitude on the part of jocks whose energies were already focused on sports rather than studies.[28] Younger teachers' more demanding academic standards, on the other hand, challenged the jocks' supremacy. Tim's recollections suggested that certain young female teachers in particular didn't appreciate the jocks:

> I had this trigonometry teacher I hated, she was young...I remember one day she told me, "You're going to be a loser." So I was—hold on a minute, so if I don't pass this class I'm going to be a loser? I won't grow up to be successful—I did this in front of the whole class—I won't grow up and be a successful math teacher like you and make thirty thousand dollars a year? From that point on we hated each other.

Tim's arrogant reaction to the teacher's bruising indictment of his abilities was not unlike how Robert Heart and his friends reacted to teachers they considered offensive. Furthermore, Tim shared a lower socioeconomic status with his predecessor of the Divided Generation, Robert Heart. Although not as poor as Robert Heart reported to have been, Tim had not counted himself among the college bound. These alumni recollections suggest that less affluent, more aggressive white males, across generations of Miller High graduates, continued to publicly assault their teachers verbally or physically. White male narrators across time also continued to justify their aggression toward school authorities by underscoring teachers' unjust or insensitive behaviors. Tim recalled how a peer whose parents had just divorced reacted to a teacher who, according to Tim, "was an idiot":

> I remember Andy Dobbs's parents split up. He was very emotionally distressed in tenth grade. The first day back to school she told him, "Just because your parents are split up don't mean you're going to get away with anything." He started shaking. He turned five shades of red, picked up the desk, and threw it at her. Almost hit her. She fell on the ground and I started laughing.

Tim's memories also echoed those of 1950s graduate Robert Heart in that Tim and his friends would step up to the academic challenge only if the relationship with the teacher was a good one. Tim recalled a particularly helpful teacher:

> She made you learn, she was actually a very good teacher. The age of the teacher was actually a big deal for me, and athletes back then did get away with things... [but younger teachers] they wanted you to learn. If they had to put up with you being a bit playful or something, as long as it did not get out of hand, sometimes they would join in.

Just like Robert Heart and his friends of the Divided Generation, Tim and his friends of the Border-Crossing Generation attended Miller High not as recipients of instruction but as critical consumers of its deliverance. For them, teachers were not professionals one was to please or receive praise from, but paid agents of a system that owed them. Thus, resisting the "system" continued to be the mark of non-college-bound young white males:

> Warren was very intelligent but he fought the system. They would fail him because they didn't like him... He was fighting the system, wouldn't do his homework, wouldn't take the tests. He was an athlete also. We both did [fight the system].

College-bound jocks Jim and Josh, on the other hand, while they too coasted through their high school years with little effort invested in their studies, enjoyed generally peaceful relationships with their teachers. Solidly middle-class white, they moved in circles where their parents socialized with their teachers and where the parent-teacher connection easily morphed from a neighborly relationship to a political alliance that favored the parents' child. They remembered:

> Most of the relationships [with teachers] were pretty good. I knew some of them... a lot of the teachers lived around here.[29]
>
> A lot of the teachers you knew through your family... your parents knew them. I think it made it easier on us... The teachers knew our parents, they had to pass us.... Like cutting a girl from the soccer team. If you knew her parents really well, you might keep her on the soccer team.[30]

While all jocks held high profiles and enjoyed special privileges, it was jocks like Jim and Josh, the privileged among the privileged, who fully belonged, and for whom the "system" was not a problem. Their recollections echoed those of white middle-class female graduates of the Divided Generation, whose relationships with teachers were intimately interwoven with their family and social lives outside of school. For white students of means across the two generations, high school was an extension of home.

All of the jocks interviewed remembered loving and respecting their coaches.[31] But white middle-class Josh White, a 1976 graduate, also remembered the coach's rougher side and the jock culture's acceptance of the coach's display of physical force as a mark of privileged attention:

> Everybody liked him [the coach], but you were kind of afraid of him too...was rough on kids too...getting beat up by the coach was like a status symbol. You were proud of it.[32]

Alumni stories further suggested that male teachers in general privileged male jocks by taking particular interest in their game wins and losses, by cheering for them, and by pitting one sport team against another. Tim Whittle explained:

> [Jocks were privileged] especially when you got the male teachers...One social studies teacher was hysterical...If he liked the sport that you were playing, you were in...On Monday, he would bring it up in class. If a kid he didn't like was on a team that lost, he made sure to bring that up [too].

Male teachers discussed the jocks' performances during classroom hours, a practice that underscored the cultural weight that sports carried among males across teacher-student boundaries at Miller High. The male bonding around sports within the school was further highlighted by contrasting recollections of female teachers' particular scorn against jocks who did not attend to their studies, as earlier discussed.

Interracial Female Jocks

Male bonding notwithstanding, female jocks, while they "fell into the jock, semipopular group,"[33] and held less status than their male equivalents, also enjoyed more privileges with teachers and a higher profile among peers than students who were not involved in sports. African American alumna Teresa Randle, a 1981 graduate, explained: "I felt that if you were on a team, they [the teachers] worked more with you, because you had to leave early a lot of times."[34] Teresa and white alumna Sophie Baker, a 1985 graduate, fully and emphatically remembered their participation in high school as jocks. In no uncertain terms, Sophie stated: "I'm definitely the jock."[35] Teresa remembered: "I did field hockey in the fall, basketball in the winter, softball in the spring."[36] In keeping with their semipopular group status, female jocks' games were always strategically scheduled as warm-up acts for male jocks' performances:

> They would place the girls' basketball games before the boys' games on Friday evening. If you saw a rise in attendance in fourth quarter, it's because they wanted to get a good seat for the boys' game. There wasn't much attendance; they just started letting the girls have cheerleaders at some of our games.[37]

Regardless of the lesser attention girls' sports drew, female athletes played, as did their male counterparts, for the sheer love of it. Similar to male jocks' recollections, and in particular to David Randle's testimonies as a black man, alumnae remembered the long-lasting bonds created on sports fields.[38] These bonds, forged on the field between African American alumna Teresa and her teammates, opened up doors to relationships with the white girls on her team that gave her some social advantages:

> I didn't drive when I was in high school. I could always call on one of the popular girls in the clique to come and pick me up, to go to a party, so I never had a problem with transportation.

These associations also exposed her to intergenerational rifts among white people at the time regarding race relations. Teresa recalled:

> I remember a friend, her father was a member of the Klan. I remember one time we were out and her car broke down. We had to go to her home. I was afraid…she said he wouldn't bother me. I was like, "Deb, I'm afraid to go into your house." "Oh, no," she said, "he's asleep. You'll be fine." I went in there with the four of us, and her mother brought us home. I was scared as anything.

Teresa's recollection of her white friend's split life, where she moved daily from the world of school and friendship with a black girl to life at home and a racist father, echoed David's and Tim's recollections of rifts among whites regarding race relations. Teresa further experienced these divisions among whites when she traveled with her white friends. She recalled:

> I used to go on vacations with my white friends. I'd go to an [ocean resort], never been to an [ocean resort] in my life. They introduced me to the [ocean resort]. You would hear little racial slurs as you were walking down the coastal highway. They were like, "I'm sorry." I didn't like it, but it was not something I was going to fight…At that time *Roots* came out. I remember going to school and everybody saying I'm sorry they treated you like that

back then... They were all very nice... Nobody treated me poorly at their home, eating dinner.

Recollections painted the world of sports as a place where students crossed racial divides and black students in particular sealed long-lasting relationships with their white peers. Black and white alumni testimonies across almost three decades of high school life since desegregation at Miller High in 1956 continued to accord sports a prominent place as where black and white students bonded within their male and female groups.

However, although in the 1970s and early 1980s many white peers felt responsible for and protective of black peers, by 1985 white students' attitudes toward their black friends were less apologetic and less constructed around the fragility of black/white relations. Sophie explained:

> The black people I disliked, it wasn't because they were black, [but] because they fell into this other group [potheads]. There was one of them that harassed girls in the hallways... He would pinch girls going by, stuff like that. We all pretty much hated him.

Throughout the 1970s and early 1980s, students' relationships across racial boundaries grew progressively more comfortable. By contrast to students of the Divided Generation, when everything "was definitely black and white," students in the 1970s and early 1980s seemed to base their feelings primarily on shared interests and values, and not on race or class. Thus Tim, a white jock, was friends with Teresa, a black female athlete, while Sophie, a white female jock who married and African American man, and her jock friends could not stand the likes of one black male pothead.

Graduates' recollections also revealed that some male jocks' notions of girls were challenging the cheerleader/jock stereotype. When one male jock remembered the athlete girls, he described them as earthy and, by contrasting them to cheerleaders, whom he ridiculed, gave them a higher status. Jock Jeremy Garnes, a 1978 graduate, remembered:

> The girl athletes back in our time, they were great. The cheerleaders, the prissy girls who didn't want to get their fingernails dirty and that kind of stuff! The girl athletes, they were just down to earth.

Jeremy's recollections of his female counterparts suggested perhaps less of a consensus among peers, and in particular among male jocks, regarding the desirability of cheerleaders. Conversely, the female jock was not completely without vanity. For Sophie, part of the allure of being a jock was the possessions that came along with the status. She fondly remembered:[39]

> I was just glad to make the varsity team because you get the jacket with the letter, with the *M* and the pins...I still have it, it's here in the house in the attic. I can't fit in it. I can't part with it. I just look at it.

Both African American alumna Teresa and white alumna Sophie "took it for granted that you needed a diploma and that it was a must." They never doubted that they would graduate. It was expected of them, and they expected it of themselves. As for the ceremonies, while Teresa reveled in them, Sophie could have done without them:

> Teresa: I was excited...It was a big accomplishment, I made it. We had a party; everybody was really excited.
> Sophie: I didn't want to go to graduation because it was this big ceremony, cap and gown...a long drawn-out calling of names, it seemed...I just wanted to go to the beach.

Although both were college bound, neither applied themselves as a student. They estimated the minimum effort required and delivered accordingly.[40] Regardless of their admitted lack of effort, neither alumna thought that Miller High provided them with a good education, echoing male counterparts' recollections. Teresa and Sophie recalled:

> Teresa: Actually, [Miller High] did not give me a good education. I think at that time, I didn't know any better.
> Sophie: "I didn't form study habits. I wasn't prepared to write on the college level."

However, they did not attribute their poor education to teachers directly, but to their own lack of effort and to the broader pervasive lack of standards that made getting by an easy thing to do, implying perhaps that a more demanding academic structure might have kept them on track.[41] Unlike their male counterparts, they assumed individual responsibility for their lackluster performances, and unlike David and Tim, both alumnae reported having good relationships with their teachers.

Of interest is that when Teresa explained her relationships with teachers, she used the term "indifference" as opposed to "different treatment," the term that had been used by African American alumni of the Divided Generation:

> In my perspective, I can't say I had any teacher treated me indifferent. That's actually from first grade on down, because I'm born and raised in this area. I went to Miller Elementary, Miller Junior High, Miller Senior High.

When African American jock David remembered teachers' relationships with black students, he too alluded to "indifferent" behavior. The earlier generation of black students at Miller High constructed cross-racial relationships more in terms of just or unjust, equal or unequal treatment of all students. Perhaps the only expression of inequality that could not elude the radar of discrimination by the 1970s was *indifference*. Unlike David, however, Teresa remembered her own relationships with teachers favorably and further attributed her good relationships with faculty to her deep roots in the community. She echoed the more affluent white jocks' recollections of life made easier by virtue of parents knowing teachers and teachers knowing generations of family members. Recollections suggested that black female athletes, already in a semiprivileged position in the school hierarchy by virtue of being athletes, might have experienced more favorable relationships with teachers than nonathletic black students in general, and nonathletic black male students in particular. Of note here is that David's recollections of teachers' indifference toward black students applied less to him than to his male black peers as he saw their athletic talents wasted. Teachers' indifference to African American non-jocks further reinforced the privileged position held by the athlete, male or female, black or white. Not unlike David, however, Teresa also recalled guidance counselors informing her only of all-black colleges and failing to mention other choices.[42]

Like David, Tim, Jim, and Josh, Teresa and Sophie remembered enjoying their high school years, which for them fully revolved around playing sports. Teresa happily recalled: "I truly enjoyed the...years I was at Miller High. I have made a lot of good friends. I still keep in touch with them. I just had a good time!"

Interracial, Cross-Class, Cross-Gender, Cross-Track Potheads

By contrast to the high-profile status held by jocks at Miller High, potheads were the "parking lot" students: "[T]hey were the people who smoked outside before they came to school."[43] Graduates' recollections revealed an animosity between potheads and jocks,[44] explained by lifestyle choices. Jocks saw themselves as healthier than potheads, suggesting that to them, drinking wasn't as bad as smoking pot. Jocks mostly "did not fool around with drugs," that was the world of the "heads."[45] Jocks drank. Jim and Josh recalled:

> Jim: "We did drink a lot."[46]
>
> Josh: "I mean, the jocks used to drink a lot...the drinking age back then was eighteen. We used to drink a lot."[47]

Jocks saw themselves as the achievers and saw the potheads as wasting their time. Josh said, "Our mailman, who was in high school the same time I was, he was a head. He admits to it. He was a mess back then. It's funny to see him now."[48] Josh's allusion to the mailman's admittance of having been a pothead at once captures present-day scorn against drug use, as well as the jock's construction of the pothead as socially inferior. Although there were also female jocks and female potheads, the rivalry polarized males of both subgroups. African American alumnus David Randle remembered:

> They [the potheads] had their group and we were the jocks... [There were rivalries] just with the jocks and the potheads.

While jocks overtly rejected potheads, as did many teachers, who called them "losers,"[49] potheads accepted anyone and everyone who shared their affinity for smoking marijuana and getting high. Within their group, they included females and males, white and black students, musicians, achieving students, and even jocks, despite the rivalry. Students were regulars or visited on occasion. For Sam Garnes, a 1974 graduate and then a regular pothead, school was an unpleasant experience. Sam, who began smoking marijuana in high school, remembered his first day as a stressful one: "I remember being confused, being worried about making it to the right classroom at the right time."[50] Unlike his jock peers who received their schedules ahead of time and became acquainted with school premises during field practices a month before school started, Sam entered the world of high school a stranger. He "didn't want to be there." His days revolved around getting stoned:

> *Did people go to class stoned?*
> All the time.
> *Did teachers react to that?*
> Nah. Not at all. Basically you'd get stoned in the morning, then at lunchtime, then after school got stoned again.

The rivalry between jocks and potheads recounted by the jock graduates was also recalled by Sam:

> There was the jocks, which were the sports people, and then there was the heads, people who smoked pot... It was the competition between the jocks and the heads... the jocks assumed they were better than everybody... and wanted to be recognized... Jocks stuck together basically. The heads hung out in the parking lot... [where] you weren't bothered by teachers.

From Sam's point of view, the jocks' arrogance divided the two groups. Beyond attitudes, the spaces claimed by each group further divided them into the more and less visible students. Jocks' visibility was pervasive. They were seen in the cafeteria, on the sports fields, and throughout school premises. Virtually all the time, they were *seen*. Not only would potheads not be seen during sport events, but they would also disappear several times a day into the outdoor parking lot, out of view. The cultural connotations surrounding the parking lot as a place where illicit behavior occurred[51] further separated the potheads from mainstream high school activities. In Sam's memory, the school was squarely divided into two groups: those who would "make it" and be successful and, on the other side, those who were considered the "losers,"[52] a concept communicated and reinforced by teachers themselves.

Throughout his high school career, Sam worked an average of twenty hours a week parking cars at restaurants, mowing lawns, and working at his father's laundry business. As was the case with many young white men of the Divided Generation, Sam visited school more than he attended it. Unlike the poorer white boys of the 1950s and 1960s, however, Sam was a middle-class suburban young man alienated from school life. Paradoxically, alienated as he might have felt, Sam believed in high school. Apart from being a place where he smoked pot with friends in the parking lot, for Sam, Miller High was also a place that anchored his life. Its usefulness resided in its predictability. If it taught him anything, it was a sense of contractual responsibility whereby he presented his body to the institution to be counted. Explaining the meaning of high school attendance, Sam said that is was "[j]ust the responsibility. Getting up and going to school, a routine for life." For Sam, academic tasks were meaningless—their rationales were not clear, and they offered no satisfaction:

> Homework, I imagine there was a reason behind that, to make you better at what you were doing, but I just didn't enjoy it.

Sam's description of homework as not being enjoyable introduced the criterion of pleasure in the construction of schoolwork. Similarly, Sam constructed teachers' efficacy in terms of their visible enjoyment or lack of enjoyment of their work as teachers, echoing other graduates' recollections of the Border-Crossing Generation regarding the importance of "liking" what one does. Sam explained:

> Some of [the teachers] seemed like they enjoyed their job, and some of them were just there to put the day in. You could sense it... they were boring, very strict discipline, no talking.

While students of the Divided Generation constructed schoolwork predominantly in terms of usefulness, and teachers' roles predominantly in terms of their acceptance and understanding of students' backgrounds, alumni of the Border-Crossing Generation added yet another dimension to their understanding of both work and teachers' roles: finding pleasure in the work itself, in the process, not just the outcome. A good teacher in Sam's opinion was "somebody that could speak to you, could relate to you...that had some different fun things [for you to do]."

Sam, a white suburban student of the 1970s, like Robert Heart, a poor rural student of the 1950s, "looked up to and respected" his shop teacher, in whose class he "was totally interested all the way." Fond memories of the shop teacher shared across generations of white male high school students who visited school more than attended it, who worked at jobs far more than they ever studied, and who, when it came to school, "didn't want to be there," suggested the importance of hands-on activities and concrete work.

Just as jocks reported that "getting beat up by the coach was like a status symbol" of which you were proud, Sam too shared memories of coaches' aggression toward young men and the respect their violence inspired.

> Shower was mandatory after gym. Back then, you used to take showers after phys. ed. Coach would smack you with the belt, whistle strap. You had respect for him. Tough, but you had respect. You would have welts, but you took it.

The image of naked bodies blistering with welts after a whipping administered in the shower stalls, a space where one is rendered fully defenseless, is likely to be suggestive within present-day perspective of abuse of power rather than respect for power. White male students' call for understanding on the part of their teachers on the one hand, and their acceptance of male school authorities' physical aggression on the other, revealed a conflicted disposition of mind and heart whereby tenderness was yearned for, but aggression was respected. This was reflected in alumni testimonies across time, and particularly those of white alumni who were marginal students. This duality—inherited, perpetuated, and transmitted across generations of non-college-bound white male students—brings into relief the emotional disconnect and physical punishment that they integrated as part of high school attendance. Making it through high school was a question of putting up with it, of doing time. While the shop teacher made it bearable for the disengaged, for non-college-bound young white men like Robert Heart in the 1950s, and Sam Garnes in the 1970s, graduation was not about skills acquired or plans for the future. It was about the end of an ordeal. On graduation day, Sam and his fellow potheads were passing a joint around in full view.

Not all pot smokers however, were considered potheads. Joanne Pet, an African American alumna and 1974 graduate, also shared Sam

Garnes' affinity for smoking marijuana. Joanne explained: "People did more drugs than they do now... the joint was pretty much the cool thing, more than drinking... 'cause the alcohol left a smell... I was pretty crazy and pretty high myself."[53] Like Sam, Joanne did not participate in extracurricular activities, and was not "a school-spirited person." Unlike Sam, however, she "loved the atmosphere of going to school." Joanne made friends easily and established longtime relationships with white female students:

> I had no transitional problem as far as making friends and getting along... My friends basically were all white, and right now, my very best friend is a white lady. Very best friend. I could call her now and she would do anything for me, and vice versa. She's not my friend because she's white, or the fact that she's not black, but her personality, *our* [speaker's emphasis] personalities.

Although she had been "pretty crazy and pretty high," Joanne "did well in school... and had no problems at all going to Miller high." Smoking pot, for Joanne, was part of the high school social scene, not an antidote to classroom boredom.

Her goal was "to be a secretary and work in an office." Although Joanne's relationships with teachers had been good and devoid of discrimination, the reality that awaited her in the job market was completely different. In this sense, Joanne's recollections echoed those of white and black women of the Divided Generation who described the contrast between great hopes and promises of high school, and the discriminatory realities of the job market:

> So I took the job... This is when discrimination slapped me right up against the head. Knocked me down like a brick wall. And then I get there, and no black people work in the office... The black people work in the factory.

In this sense, for women in particular, the middle-class white women of the Divided Generation and the black middle-class women of the Border-Crossing Generation, high school continued "to provide... a social experience that [was] markedly more egalitarian and more open to free choice and possibilities of self-realization than anything that [was] available to them in the realm of work."[54] Graduating from high school, however, was a time of great pride for Joanne:

> That was a big thing. Had a graduation party... My brother came home from Ohio... and he drove home for my graduation, him and his family,

and that was the highlight for me, 'cause he didn't come home often...He and I were the only ones who graduated from high school.

Sharing her achievement with the only sibling who graduated from high school was deeply meaningful to Joanne, who grew up in a family that paid little attention to acquiring an education. She remembered the messages she received in early childhood:

> When I was a kid growing up in my house, it was kind of cut and dry. "Oh, you didn't go to school today?" "Nah, tomorrow." "Hum. Didn't go to school today?" "No, later." "Well, maybe you need to look for a job."...It was just like, "I don't want to go to school anymore." "Okay. Get a job."

Joanne's recollections challenged the stereotype of the pot smoker as "loser" on several levels. Not only did she succeed in acquiring a high school diploma against the odds in view of her family background, but she also did not lose track of her desire to become a secretary, even though she regularly hung out with pot smokers. Her ability to make friends easily notwithstanding, Joanne explained her high school success, and indeed the rewarding life that she has led, in terms of learning from others' mistakes and poor choices. Joanne defined the term mentor in an unusual way:

> A lot of people today look at mentors to be great people—Michael Jordan, Oprah Winfrey, whoever. My theory is it doesn't have to be that way. Your mentor can be the other direction. My sister, she got pregnant she was eighteen, got married, and the next day had another baby, had another baby, bad marriage, low income. That was my mentor. Something *I do not want to do* [speaker's emphasis].

While graduation was a time of great pride, it was not a time of hope for Joanne, as it had not been for many black youth of the Divided Generation. She recalled: "I didn't see any hope. I guess the only thing I knew was you graduated from high school and you went to work." Joanne's remembered lack of hope during graduation but pride in the act of graduating revealed a mixture of stoicism, akin to that of working-class white young men proud to have made it through the system, and of disappointment, akin to that of white women of the Divided Generation whose experiences of success within high school would not be matched in the job market. Thus, young women and men, white and black, for whom college was not an option, as alumni recollections revealed across the decades and into the 1970s, experienced high school as a personal challenge. The diploma was the symbol of beating the system, or vindicating parents' background of poverty, or dissociating oneself from a life of bad marriages and ongoing pregnancies. The diploma, for these students, symbolized not so much what they had achieved but what they had avoided, and perhaps in that sense did not offer much hope. The diploma attested to the fact that they were not failures rather than that they could be successes.

Interracial Male Musicians

Beyond jocks, potheads, and friends of potheads, some students identified themselves as musicians. This was the case of African American alumnus Pat Baley and white alumnus Michael Hallner, graduates of 1975 and 1976.[55] While in popular culture, musicians in the 1970s were stereotyped as drug users on a perpetual high, and while it is almost certain that many Miller High musicians were pot smokers, Pat and Michael were not "airheads." However, they shared with potheads such as Sam a peripheral involvement in school. Pat explained: "We [musicians] didn't try hard, we just got by. Michael echoed: "If I got Cs, that was good. The goal was to get through it, get done with high school."

For middle-class African American alumnus Pat and middle-class white alumnus Michael, playing music was more important than anything school had to offer. Getting by with the anticipation of getting out was how these garage musicians made it through high school. Their lives began after school. Michael remembered:

> When school was out, it was my time and that's how I felt about it. So it's time to go home and do my thing, you know.

The world of musicians that Pat and Michael recalled was solely a male world. In Pat's band, males were black and white. In Michael's, they were all white. Unlike male and female jocks and unlike Joanne Pet, adolescents who looked forward to going to school, Pat and Michael looked forward to leaving school and playing music in their garages. Their recollections echoed those of many white and black young men of the Divided Generation. To them, schoolwork seemed irrelevant to life at large—"It was dull."

For Pat and Michael, the "goal was to get through. It wasn't to excel, it was: you're here, just get through it, get done with schoolwork, get on." Like many young males, black and white, who stuck it through, neither considered dropping out. Michael never thought of dropping out for fear of having to contend with his parents' displeasure. Typing was the only class he remembered as being of some use to him in later life: "I learned more in that class that benefits me now than anything. Learning how to type was the best thing."

While Pat also just "got by," he saw the usefulness of a high school education in terms of social capital and the relationships that turned out to be important connections for economic success later in life. He explained:

> You got social skills [at high school]. You met people there that... down the road might own a business and you got a job 'cause you knew them.

Pat and Michael were both peripheral students who put little effort into their studies, did not participate in any extracurricular activities, and lived to play music after school hours. While for Michael, a middle-class suburban white young man, the whole experience was fairly useless, for Pat, an African American middle-class young man, high school provided the opportunity to seal friendships across racial barriers, friendships that proved to be useful connections later in life. For white young males of some means who graduated in the 1970s and early 1980s, as had been the case since the early decades of the twentieth century, high school had little bearing on their economic fortunes. Furthermore, the particular demographics and geographical configurations of Miller Town, which more readily juxtaposed richer and poorer, white and black, gave many black families in the neighborhood the opportunity to live side by side with wealthier white people. Pat recalled: "I grew up in a neighborhood where there was nothing but [white] rich people. Now I wasn't. But they were my best friends."

Other Interracial Peer Groups and New Attitudes across Race toward Pregnant Youth

Recollections brought to view other student-generated labels that identified the very smart and academically successful male students as "eggheads"; the mechanically inclined shop students, all males, whose passions were to work on cars as "motorheads;" and the "fireheads," all males again, who volunteered for the town's fire department and spent as much time as they could around the firehouse. Except for the labels "jock," "prep," and "pothead," labels were generally assigned to males: "The guys were eggheads, girls were just smart."[56]

Although young women continued to get pregnant at Miller High, they were less likely to leave school than pregnant students in the 1950s and 1960s. African American alumna Teresa explained:

> There was a pregnant girl, a white girl, she had the baby, gave it up for adoption, and came back to school. Another girl had a baby, kept her baby, and

came back to school, so she was a mother when she was in junior high; she was black. Then a couple of girls that walked across the stage on graduation that were pregnant. They ended up having the babies right after graduation from high school.

The Poor and Other Marginalized Students

Only two alumni remembered "poor" students whom they described as marginalized. Musician Michael Hallner recalled:

> The few that were poor, they turned out to be the ones that were picked on, not because they were poor, but the way they dressed. There was one girl, one boy... they were the same ones all the way through twelfth grade... almost every kid in the class would pick on them... In twelfth grade the girl was there, but I don't think the guy was still there.

Tim Whittle also remembered, but more harshly:

> The real poor kids were teased just because they were dirty... they didn't get teased because they were poor, they got teased because they were pigs. They kind of brought it on themselves.

Tim's particularly virulent indictment of the poor might have been a reflection of his own need to distance himself from poverty that lurked around the corner as he was growing up. Michael's and Tim's recollections depicted merciless behavior toward the poor identified also in recollections of graduates of the Divided Generation, suggesting a continuation into the 1970s and 1980s of the disdain for the poor, whose "dirty" appearance betrayed their economic status. Except for Michael's and Tim's testimonies, other graduates' recollections during this period did not bring into view the poor, even when directly asked about them. In fact, African American alumna Teresa, who graduated the same year as her white peer Tim, remembered only middle-class students: "Everybody was middle class. One or two working class, everybody was in the same class."

This generation's more relaxed appearance might have loosened the hold the clean look had during the Divided Generation, and might have made the "sloppy" look more acceptable, blurring lines between haves and have-nots. Alumna Joanne Pet described potheads in this way: "[Potheads] had really, really long hair, raggedy clothes, eyes red." However, the teasing of the very poor as reported by musician Michael and jock Tim might have been perpetuated by enough students for the two of them to have noticed and remembered. The very poor at Miller High would continue to be stigmatized as personally deficient; and, as in the 1950s and 1960s, the very poor would continue to drop out, as Michael's testimony quoted earlier suggested.

Another group of marginalized students came to view indirectly, as remembered by others. While these students did not count among the very poor, they were not going to college, did not participate in sports, and were part of the working class, as Tim Whittle recalled. They were treated differently, not by students, but by teachers.

> "I saw being treated differently were... your group that did work-release, that took half-days and went to work. It sounds like they were in prison. I didn't like the word "work-release." Most of these guys didn't play a sport.... Guys who weren't connected to school... and didn't want to be involved in school. Some of the teachers gave up on them, on certain kids."

These students lived their lives within a school sanctioned limbo-like structure, where they were neither fully students nor fully employees. The work-release students were strangers to the school and its students since they attended only half-days. They were denied the social interactions that, as alumni stories have consistently revealed, offered opportunities for amassing social capital, for practicing alternative identities, for imagining a future different from one's past.

The Preps

Of all the peer groups identified by alumni of the Border-Crossing Generation, one group of students, the "preps," continued to hold a highly visible profile. They produced the yearbooks, wrote for the newspapers, organized school dances, and continued to create the image of the school. Described through the eyes of African American alumna Teresa Randle and white alumna Sophie Baker, the preps of the Border-Crossing Generation resembled those that Linda Moss, alumna of the Divided Generation, described as "the doers... the people that were on the teams... were the people that were putting out the yearbook, were the people doing that, you know, class officers...."[57] African American jock Teresa Randle remembered the preps as "the ones that were the class officers, cheerleaders."

However, Sophie Baker's description of the preps is the first among alumni recollections to represent the "doers" in an unfavorable light. Sophie remembered the undemocratic actions of the preps, who wore "alligator shirts and docksiders," and how they imposed their choice of prom song against majority will:

> They were... in the SGA. They sent out ballots for what you wanted to be your senior prom song. It seemed everything we voted for never went. We would count how many people wrote [in the ballot] this song [that they chose]... how did this [other] song make it?

By the 1990s, as we shall see, many more voices would criticize the preps and boycott their orchestrated school activities.

A Border-Crossing Generation Pushing against Structural Forces

Rivalries among peer groups notwithstanding, the Border-Crossing Generation of the 1970s and early 1980s ventured more freely across class, gender, and race divides, and tended to submerge these categories under youth-generated identities, from potheads to musicians. This was overall an accepting generation. African American jock Teresa Randle remembered:

> Everybody was allowed to be their own individual self. I don't remember anybody being aggressive toward anybody. They just knew that they hung over here, and the other ones hung over there. It was a given.

However, while students of the Border-Crossing Generation interacted more democratically than had the previous generation, many still contended with structural forces that constricted their movements across racial and class categories, even as peer groups offered opportunities for trumping race, class, and gender, as I explore in chapter 4. In the chapter that follows, I also examine the changing meanings of race for white youth, and the negotiations of race by black youth.

Chapter 4

Bridging Social Divides Through Peer-Groups: A Socially Tolerant but Politically Inactive Student Body (1970–1985)

Miller High students, as earlier suggested, associated mostly on the basis of common interests and affinities within interracial, cross-track, and sometimes even cross-gender peer-generated groups. Describing the cafeteria, Tim Whittle recalled that "male jocks [black and white] had a table, girl jocks [black and white] had a table, the eggheads [black and white] had a table...," and so on. Overwhelmingly, graduates remembered blacks and whites getting along within and across the peer groups.[1] African American Joanne Pet's recollection captured the civil rights mood of the time:

> Given that it was five or six years after Martin Luther King and the civil rights [movement], I was right there on the crest of it. So I didn't feel the effect of racism...I don't remember any incident that I felt that I was discriminated against because I was black...My friends, basically, they were all white.

However, structural forces—namely a split, inconsistent, and at times grossly insensitive faculty—divided students into winners and losers, neglected whole segments of students, and, as we shall see, crossed boundaries of propriety with their charges. Additionally, a continually pressing hierarchy of academic tracks continued to constrict the range of possible relationships of Miller High students even as they expanded their relationships across racial, gender, and track divides. Structural forces notwithstanding, within the setting of an increasingly suburban town and a national atmosphere of expanding civil rights, the Border-Crossing Generation interacted more democratically than the previous generation, although it never directly protested the status quo.

Peer Relationships

New Constructions of Race Relations and the Role of Peer Groups

From the majority of white students' perspectives, as suggested by recollection of both white and black graduates, outsiders were not the black students, but the minority white racist students whose discriminatory behaviors would quickly be brought to a halt by white peers. Unlike the Divided Generation, when black students who were assaulted by white peers sought help from school authorities, black students of the Border-Crossing Generation found ready help from most of their white peers. There seemed to be a more deeply integrated sense of racial equality among white students of the Border-Crossing Generation. This deepened sense of racial equality on part of the majority of white youth in the 1970s was further evidenced in the schisms in racial attitudes within white families, as the new generation of whites sought friendships with blacks against parental wishes.[2]

Graduates' testimonies further suggested that peer-group affiliation counteracted broader societal inequities and discrimination since belonging to a peer group typically meant belonging to an interracial group. However, students who did not belong to a peer group continued to segregate by race and class. Membership in a peer group offered an identity to layer over raw race and class distinctions. Without the cloak of peer-group appurtenance, one remained exposed as a census data category.[3] As earlier mentioned, using the phrase "social class" to describe peer groups, Tim Whittle recalled:

> If you weren't in a social class and you were black, you sat with the black kids. Teresa sat with the girl athletes; Amy sat with the girl athletes. Rona sat with the black girls because she didn't play a sport.

In the cafeteria, black students who were not part of the jock group or any other peer group sat together as black students. Of course, the story is more complex. African American alumna Joanne Pet explained that it was "a comfortable type of segregation":

> You segregate yourself regardless. We as people do it today. It's just a natural thing... black people sitting over here, guys over here, girls over here... It's just a natural thing.

Joanne's recollection about black students sitting together at lunch out of comfort rather than because of racist attitudes on the part of white students was also echoed by African American alumnus David Randle, who, while he felt at ease as a jock "eating anywhere," remembered the cafeteria segregation as a question of feeling "comfortable":

> We had open lunch back then, so you could bring in music. So naturally we [the blacks] wanted to hear something, and they [the whites] wanted to hear something else. They didn't force us in any groups. Just felt more comfortable.

White alumnus Time Whittle also remembered what happened when one didn't belong to a peer group: "You went with who you were most comfortable with... you went with your comfort area, kids you knew when you were younger."

Thus, a segment of the African American student population was primarily identified as black by their peers, rather than as jocks, musicians, potheads, fireheads, and so on. Many black students kept to themselves, whether proactively by refusing to participate in the white culture, as David's recollections suggested, or by default, either because they were not recruited by coaches into the mainstream sports culture, or because, as in the case of many young black women who continued the previous generations' roles of caretakers at home, they could not participate in after-school social activities with their peers. This segregation of black students suggests perhaps that the idea that everyone intermingled was the construction of the privileged students, white and black, whose high profiles at school, and/or identifiable appurtenance to a peer group, made it easier to navigate social borders of race, class, and, in some cases, gender. Lack of opportunity to participate or join peer groups and the natural tendency to seek comfort with one's own resulted in segregation by race for a segment of the black youth.

But "work-release" students also segregated. Work-release students were usually youth of lesser means who belonged neither fully to the marketplace nor to their school, and who were looked down upon by teachers, as we shall see in greater detail. The segregation of some black students and of work-release students restricted their opportunities to establish or broaden social connections. African American male and female jocks David and Teresa, African American male musician Pat Baley, and African American alumna Joanne Pet expanded their social networks within interracial peer groups, and forged interracial connections that continued into adulthood. Lacking peer-group affiliation, one became, by default, a member of communities that were racially and socioeconomically defined.

Overall, during the 1970s and early 1980s, male and female, white and black Miller High students along the upper- to lower-middle-class continuum experienced greater freedom to be themselves with each other, regardless of skin color and gender, than graduates of the Divided Generation ever had. Teresa Randle's testimony, however, of a visit to the house of a white friend whose father was a member of the KKK suggests a disproportionate burden placed on African American youth in relating across racial boundaries. Her testimony reveals the nonchalance of the white girl who seemed oblivious, perhaps by virtue of a combination of privilege and adolescent bravado in face of the status quo, to the terrifying implications for a black young woman of a possible encounter with a member of the KKK. It equally underscores the willingness and ease with which a white youth of the Border-Crossing Generation challenged the status quo, but again, with disproportionately little discomfort to her "self" in comparison to her black friend.

New Constructions of Gender Relations across Race and the Role of Peer Groups

By the 1970s, the dance floor was no longer a threatening place for black students. In contrast to the 1950s, when white middle-class teachers tacitly supported white students' exclusion of blacks at school dances, the 1970s and early 1980s witnessed an emancipation of allowable relationships on the dance floor at Miller High. Not only had it become a place where white and black couples danced side by side, but also a place where black males danced with white females and white males danced with black females. African American alumna Joanne Pet remembered:

> I had no problems looking at the white boys, thinking they were cute or vice versa. I had this guy...and the boy was drop-dead gorgeous, blond-haired white boy. And he asked me to parties. I mean it wasn't a question of am I going with him. What else am I going to do? If I'm living, I'm going. I'd have to be dead not to. You know what I mean?

Speaking about interracial couples, African American jock Teresa remembered: "The junior class, there was a couple. I can still see her, I can still see him...they are still married."

While students of the Border-Crossing Generation had substantially democratized dating patterns when compared to the Divided Generation, and interracial cross-gender relationships were now acceptable, new relational hierarchies emerged at Miller High, the result of the introduction of football at the very end of the 1960s. As several recollections revealed, jocks dated the cheerleaders and the best-looking girls in school. By the 1970s, cheerleaders were black and white. However, jock Jim Garnes' testimony about his preference for female jocks over the cheerleaders, addressed earlier in chapter 3, challenged the stereotype that male jocks are only interested in cheerleaders. Jim's memories of his female counterparts suggested less of a consensus among peers, and in particular among male jocks, regarding the desirability of the "cheerleader type," further suggesting that inherited binary constructions in popular culture around jock/cheerleader, popular/unpopular, or attractive/unattractive might never have represented the lived experiences of most high school students, but were the highly visible experiences of a very few. It might not be too bold to advance that "the distorted lens of historical mythologizing"[4] might be at play as movie directors, advertisers, and crafters of popular culture perpetuated and perpetuate certain stereotypes in high school peer associations.

Recollections also suggested that young women, when compared to young men, continued to be perceived as "definitely...a lot smarter."[5] From the 1950s to the early 1980s at Miller High, peers, and in particular male peers, whether white or black, continued to perceive females as smarter and better students. Female intelligence was also linked, across time periods, to effort, evidenced in graduates' references to young women doing homework for their male counterparts. The feminization of the "good" student throughout these periods, a characterization that shifted into the less feminine and more "purposeful" student in the 1990s, as we shall see, suggested an understanding of good students as those who did as they were told. Something that, at least as reported by graduates interviewed across the two generations, was better accomplished by young women than young men.

Youth and Authority

Constructing Teacher-Student Relationships across Social Categories

Unlike the Divided Generation, the Border-Crossing Generation depicted a student body less involved with faculty and school authority. For one, seldom did graduates, across race and gender, spontaneously remember their principals, and when asked about them, they vaguely recalled their presence, or not at all.[6]

Three principals succeeded each other between 1970 and 1985. The first two principals held office during 1970–1973 and 1973–1978. Although the third principal, Mr. Krauss, held office much longer, from 1978 to 1991, he was not remembered fondly by alumni who graduated before 1985. The very few graduates who remembered Mr. Krauss as an effective principal were graduates of the early 1990s, as we shall see, and for reasons connected to increased school violence at the end of the century. Recollections of the Border-Crossing Generation revealed an administration uninvolved in students' daily lives except when punishment was to be administered. Jim Garnes recalled:

> They [principal and vice principal] were distant. They didn't walk the hallways or anything... Only time you saw a vice principal is when you were in a lot of trouble. You were going to get yelled at. If you ever saw the principal, then you were in real, really big trouble. You really did something wrong.

While Mr. Lancaster had been beloved by Miller High students throughout his 1949 to 1964 tenure (and from 1956 on by both black and white students), as had been his successors who served during 1964–1966 and 1966–1970, principals in the 1970s and early 1980s were perceived by Miller High students not only as distant and punishing, but also as vengeful. Tim remembered: "This one vice-principal wanted [a buddy of mine, a football player] out of school."

Students' relationships with teachers, on the other hand, were not as clear-cut, and continued unevenly, as they had been for the Divided Generation. Students who butted heads the most with faculty were non-college-bound white males of lesser means, as Tim's recollections revealed. Some of these young men who either avoided or fought with teachers established connections with a coach if they were involved in sports (like jocks David and Tim), or with the shop teacher (like pothead Sam Garnes). The persona of the coach loomed large in the memories of black and white males for whom sports provided a place to belong. Some males in the 1970s seemed to accept and find status from being "beat up" by the coach, whether these young men were jocks or potheads. There is reason to believe that the introduction of football jock culture, nonexistent at Miller High before the very end of the 1960s, might have contributed to the greater acceptance on the part of many male students of coaches' aggressive behavior toward them as part of a manly show of stoic endurance.[7]

While coaches beat up on young males, other teachers literally made love to their female students. Narrators' recollections of teacher/student affairs in the 1970s and early 1980s brought into relief the transformation of a high school climate with the infusion of a very young faculty. Reports of student/teacher affairs that emerged in recollections of alumni of the Border-Crossing Generation were also remembered by graduates of the 1990s, suggesting a continued trend in teacher/student affairs for at least the latter three decades of the twentieth century at Miller High.

Sophie Baker recounted an affair between her friend and their teacher, an affair that escaped the radar screen of both school authorities and parents. Her recollection underscored not only the young friend's inner turmoil and vulnerability, but also the young women's fascination with carrying out a love affair with an older man, a fascination that helped Sophie be an accomplice to the illicit relationship. In hindsight, at the time of the interview, and from the current perspective of a mother of a teenage daughter, Sophie regretted her participation in her friend's affair. As a youth, however, she vicariously lived out a fantasy through her friend's involvement with the older male teacher, and intensified the experience by keeping it a secret at the request of the teacher.

> I'm embarrassed to say this. I helped her. He was younger, closer to our age, and when teachers...say they care about you, love you, and she was having problems at home, parents fought all the time, as an adult I can look back and see that she needed attention, and he gave her attention in the wrong way...I was a little bit enchanted with the whole older man thing...I would drive her to his van...he made us swear to secrecy...This girl had low self-esteem, and when you take advantage of that, that makes you a predator.

Tim Whittle also reported rumors of an affair between an African American teacher and a white student in the late 1970s that, although unconfirmed, sketched out the pervasiveness in young people's imagination of the time of the possibilities of teacher/student love affairs. Tim recalled:

> I got to Miller High a year late on a scandal with a teacher and a girl...I think '77. A black male teacher with a student white girl—that was a double thing. I don't know if she was pregnant or if he just slept with her. That was some kind of a scandal...he disappeared.

Evidence suggests that the influx at Miller High during the 1970s and early 1980s of younger teachers, who, closer in age to their charges, spent time with students outside of school hours and fraternized with them, created a social environment where teacher and student could readily become friends and lovers. This might have been further supported by the broader social and legal climate of the times where students' rights were on the rise and school authorities' paternalistic roles on the decline.[8] Sophie recalled:

> When I was in school... I spent time with people [school peers and teachers] on the weekends, it was platonic [with] the teacher... especially with sports, [spent time] at [teacher's] house, cookouts.

Young teachers, as alumni reported, were also more apt to put up with student misbehavior if it meant greater student participation in classroom discussions, while impatient older teachers routinely sent their charges to the principal's office.[9] These younger teachers not only loosened rules of engagement within the classroom, but also warned students ahead of time when administrators were displeased with their behavior. In essence, teachers snitched on more senior school authorities and acted more as peers to their students than authority figures. As one alumnus put it: "the younger teachers would tell you when someone was on the prowl for us," while the old teachers were more likely to "send you to the office."

Young teachers' disengagement from the establishment on the one hand, and their behavioral allowances on the other, contributed to blurring the traditional teacher-student boundaries for both teachers and students, in turn making student-teacher affairs easier.

However, the more empowering consequences one might have expected from greater closeness with teachers, such as greater participation on the part of black students who did not belong to peer groups and on the part of work-release students, and in general a more politicized student body, never materialized. The split between much younger and much older faculty, according to recollections, created a rift that focused teachers' attention more on other teachers than on students. Tim recalled that there was "a huge difference" between the young and old teachers. He said: "I don't think they liked each other. That is what cracked me up."

While in rare cases female students were reported to have affairs with male teachers, in general, female students' relationships with their teachers were far less disappointing or contentious than they were for male students who were not college bound. Alumnae in general reported good relationships with teachers.[10] By contrast, alumni fondly remembered mostly the coach and the shop teacher. For nonintellectual males, who constructed their expertise in terms of physical prowess, discipline, and teamwork in sports, or by designing, organizing, and producing projects in shop—hands-on activities that required physical and manual dexterity—the coach and the shop teacher were influential male role models. However, while the coach enjoyed relatively high status at Miller High, the shop teacher held the lowest position on the hierarchical totem pole of teachers and subject matter. Whatever satisfaction male students might have derived from building and producing concrete objects in shop, one might assume they also integrated the institutionalized low value that such activities held.

Male students who took shop or were part of the work-release program held a lower status in the minds of teachers. Teachers "liked the smart kids, and the jocks."[11] Miller High faculty communicated a bias against manual work and part-time school attendance through work-release programs. Students themselves, however, did not share in their teachers' bias, strongly suggesting that the privileging of certain types of knowledge-making sites (i.e., shop versus math class) was an adult construction no longer espoused by high school students in the 1970s and early 1980s, as it had been by students of the Divided Generation.[12] Teachers, on the other hand, arranged students in polarized fashion, either vertically, as losers or winners—"A trigonometry teacher... [she said,] 'You're going to be a loser'"[13]—or as insiders and outsiders, as in the case of work-release students.

Examining Miller High from student perspectives brings into question the institutionalized hierarchy of knowledge-making sites—courses that are offered but that are not equally regarded. For example, teachers looked down on shop when compared to the academic track. For many male students across class and race, from the 1950s to the early 1980s, learning was meaningful when they were engaged in manual work, or in making music. Yet many teachers devalued these learning spaces. Even when school authorities officially sanctioned an activity deeply rooted in a social-efficiency goal[14] such as the work-release program, teachers devalued participation in the program, de facto devaluing it as a legitimate site of learning for students.

The absent principals, the disinterested older teachers, the imprudent younger teachers who snitched on administrators, and the feuds among faculty along the generation gap combined to weave an inconsistent and often unresponsive environment for students who did not count among the smart and the jocks. Thus, students of older teachers did well if they behaved, whether they studied or not. Students of younger female teachers did well if they were smart, whether they behaved or not. Male jock students of sports-minded younger male teachers did well even if they weren't smart. One can only imagine the myriad contradictory messages that young people sorted through and around which they negotiated their daily student lives. One can also imagine how such adult behavior would create in the eyes of young people a laughable high school world at best— "that is what cracked me up"—and at worst an alienating world in which they perceived that teachers "gave up on you."

Ironically, teachers accorded young people varying degrees of legitimacy as high school students depending on their peer-group affiliations. The affective importance attributed to group belonging, by students and teachers, is a phenomenon perhaps unique to the U.S. high school—a phenomenon that paradoxically speaks more of individuality than of community or group belonging.[15]

Students of the Border-Crossing Generation who did not belong to peer groups acquired racialized and class identities. Teachers, as recollections suggested, would lose sight of students as individuals when those students did not belong to a particular peer group. Thus, black students who did not belong to a peer group were robbed of "individual" status. They became the black kids who, as David's testimonies suggested, went individually unnoticed by the white faculty, something he lamented in view of the many talents he knew many of them to possess. David, on the other hand, by virtue of being a jock, was accorded individual attention. Work-release students who lived in a limbo state, neither full-time students nor full-time employees, were also robbed of individuality and were seen as the working-class people coming and going at the periphery of school life, a group whom many teachers "gave up on." This phenomenon, by which specific peer-group affiliations granted individual status and lack of specific peer-group affiliations erased the individual within racialized and class groups, did not automatically equate individual recognition with positive recognition. Potheads were usually identified as losers by teachers, as Tim Whittle recalled: "A buddy of mine... he loved to get high, the teacher said, 'I'm going to make sure you don't graduate high school, you're a loser.'" But it did accord individual attention.

Unlike recollections of black graduates of the Divided Generation, whose overall impression of school authority was favorable while relationships with peers were reported as problematic, black graduates of the Border-Crossing Generation experienced the reverse: a rapprochement with students, but a greater alienation from high school authorities. Nat Right, the African American alumnus of the early 1960s who was assaulted by a peer, remembered his principal fondly as a fair man and spoke of students finally getting to the "business of being students." By contrast, David, an African American graduate of the 1970s, remembered poor and disappointing relationships with teachers and administration, but very good, long-lasting relationships with white friends. In the 1970s and 1980s, "resisting the system"—which had been the modus operandi of many poor white males during the early days of desegregation at Miller High—was now becoming the modus operandi for black males who were not jocks.

Young black men to whom faculty and administration had not reached out and who could not find a space among peer-generated groups retreated into voluntary racial/cultural segregation within the comprehensive high school. They resisted the white system, as David's recollections revealed. This segregation, as earlier discussed, was not aimed at white students, but at the white school authorities. On the other hand, as yearbook data show, more young black women participated in school at all levels, and as discussed earlier, had better relationships with teachers in general.

Students' Social Liberalization does not Translate into Political Action

There is reason to believe that youth-generated peer associations within a generally tolerant student body where "everybody was allowed to be their own individual self" created social conditions within which peers were less likely to compare each other based on the academic tracks they attended. Why, then, did there not emerge a student-generated change in status allocation of subject matters or in content of subject matters? Not for the sake of quickly tucking away credits toward an easy credential, as education scholars have documented,[16] but for the sake of surviving an otherwise meaningless environment, as in the case of Sam and others like him for whom shop, for example, was a site of meaningful learning. Miller High students did not take the opportunity of their particular space in time to challenge adult-communicated hierarchies of learning sites, and instead retreated from confrontation with school authorities into their bounded or hybrid communities. Whether these were black non-jock males who refused to join the system, black and white musicians who lived for their music, or potheads who found solace in shop class, the opportunities for protest were there for students. High school students in other parts of the United States during the late 1960s and early 1970s, and in urban settings in particular, had already begun and continued to articulate their dissent against institutionalized inequalities. One such white urban student deconstructed the high school system as a reproduction theorist might have and articulated a desire for unity among youth across racial, class, and track divides.

> High school is used to put people in various slots. It puts black people or poor people into slots—they will be working class. People like us here will go to college and flounder around in the arts. We are not supposed to have any consciousness of what goes on in working kids' minds...kids are coming

to realize one another's needs and want to break out of this classification system.[17]

Similarly, African American alumnus David's disappointment with the only black teacher, who taught history, echoed that of city-dwelling African American youth of the earlier twentieth century, as documented by historian E.F. Frazier.[18] In *Negro Youth at the Crossroads*, Frazier quotes a disgruntled black youth whose feelings echo those of David: "I've often wondered why we didn't study more about Booker T. Washington than George Washington! No matter how much I try, I can never be George Washington." David's disappointment underscored the historically known dissatisfaction of black high school students with the white bias in the history curriculum and simultaneously revealed the absence of overt complaint by Miller High black students in the 1970s against such biases.[19] Young black men of Miller High who did not participate in sports and who were not academically engaged asserted their dissatisfaction by not participating in school life rather than protesting its curriculum.

There is reason to believe that the school's relatively small African American student population in the 1970s, the close-knit community where long-established black and white families shared an identity as Miller Town residents and within which African American small-business owners counted on white clients, and the one black teacher's reluctance or inability to galvanize the energies of Miller High black male youth toward academic participation combined to silence dissatisfied black youth whose families emphasized respect for authority and getting along with everyone. African American alumna Joanne Pet summarized other African American graduates' recollections of their upbringing: "I was familiar with getting along with white people."[20]

The lack of outreach on part of the white faculty to recruit black youth into sports or extracurricular activities, the one black teacher's failure to deliver the African American historical reality that might have motivated and inspired Miller High's black student population, and the force of parental influence that emphasized "getting along" contributed to keeping many black youth covertly dissatisfied and seeking solace in group rejection of a system by which they felt rejected. This was, however, a group rejection manifested in nonparticipation in extracurricular activities and in skipping classes, but not in dropping out completely or overtly protesting change.

However, while Miller High students, white and black, did not make public their possible critiques of institutional norms, they seemed to ignore adult-imposed academic and social hierarchies. Miller High students' attitudes toward various sites of knowledge making (i.e., classrooms, shop, sports field, work-release, etc.) were liberal and democratic, while teachers' attitudes were elitist. Overall, students arranged themselves on a horizontal continuum that juxtaposed them by labels that captured their passions and interests, rather than on a vertical continuum that divided them by scholastic or extracurricular accomplishments, even as teachers favored jocks and the smart kids.

Still, in reaction to a generally removed and unresponsive white administration and faculty, some black students abstracted faculty and administration into an all-encompassing and nebulous "white system," just as they might have felt abstracted as individuals into the overarching "black students." This was similar to the way that white males of lower socioeconomic backgrounds of the Divided Generation abstracted high school life into "the system," as they too felt themselves unseen and unrecognized.

The result going forward would be a hardened identification with race for many black students, and a hardened identification with the lower class for many white work-release students. These hardened divisions, as we shall explore in Part III, provided the foundation for a complete reorganization of student associations. The affect that sprung out of those hardened identifications, coupled with an explosive demographic shift and zero-tolerance policies delivered by increasingly distant school authorities, created the conditions for a rigid resegregation among students.

Part III

The Redivided Generation (1986–2000)

Chapter 5

Memories of Segregation by Class, Race, Nationality, and Religion: Destabilizing Years of Shifting Demographics (1986–2000)

In the late 1980s and throughout the 1990s, Miller High students attended an overcrowded high school where it was "tough to get from one end to another [between classes]...If your class was next door, literally you had to fight through the hallway."[1] It was a school that by one alumna's recollection was "old and ugly."[2] The solution to the overcrowding came only at the very end of the century with the construction of a state-of-the-art 600-student addition that was completed in 2000.[3] An influx of new arrivals to Miller Town contributed to the overcrowding: immigrant Russian Jews[4] and ex-city dwellers, many of whom were taking advantage of the recently erected Section 8 apartments[5] and most of whom were black. During this period, the population explosion accelerated the expansion of Miller Town, as high-end and low-end houses filled up any remaining farm fields, restaurants multiplied, and grocery chains, gas stations, antique shops, shopping centers, and malls spread.[6]

This accelerated development and avalanche of newcomers created a reaction from both whites and blacks with long-time roots in the community toward all those whom they called "newcomers" and accused of importing bad city habits. Still, while a young Russian immigrant might hear an irate Miller High peer shout, "Russian go home,"[7] irritations by old-timers were mostly directed toward ex-city blacks who doubled the population of black students at Miller High.[8] Graduates explained the changes in Miller Town population in the 1990s:

> By the 1990s, the black population in Miller Town had changed a lot, because I don't like to say anything bad about it, but look at all the apartments... We didn't know those people, though. They were all newcomers.[9]
>
> In the '90s, that's when blacks that weren't raised here started moving here... This area just exploded. In the '90s, the population probably tripled. Farms were dropping left and right. Black kids from other towns were moving here, and they were different from Miller Town black kids, totally different. You ask the black kids from Miller Town, they will tell you. I will quote you my buddy C.J. [black friend], he said to me, any time a county black starts hanging out with a black from the city, the black kid is either dead or in jail. We have friends that are dead from high school because they started getting hooked up with other black kids from the city, then they became heroin addicts. I'm not saying it's their fault; it's just a different culture. Every kid I grew up with that was black, soon as he started hanging out with non-county people, black or white, they were in jail or dead... you're mixing two different cultures... that's when they had some racial issues... black kids from the city.[10]

In response to what many perceived to be city kids' nefarious influences on local kids, the blacks and whites who grew up in Miller Town and were now having families began sending their children to private schools or homeschooling them.[11]

In the last decade of the twentieth century, Miller Town transformed from a place where one was likely to cross paths with a known face into a town of xenophobic old-timers and eager newcomers hoping to find a better life. The town's high school was absorbing an unprecedented amount of newly transplanted youth "with attitude," among which a segment of the recently immigrated Russian youth, who identified themselves as the Russian mafia, trafficked in drugs, hated blacks and rednecks, and sported anti-American sentiments. "Rednecks" in turn, who as previous analyses revealed, had consistently displayed racist attitudes toward blacks, were no longer being silenced by vigilant white peers as they had been during the 1970s and early 1980s. To their list of black foes, those identified as rednecks now added Russian Jews, as we shall further explore.

No longer was Miller High nestled within an out-of-the-way rural-suburban town. It was now the high school of an urban-suburban town easily connected by a metro line to downtown city life. Miller Town had become accessible to visitors who could not afford to own cars. Black parents in the city desperate to see their children attend the better county schools concocted fictitious residential addresses and began sending their youth to Miller High, to the growing frustration of administrators who began investigating the legitimacy of black students' school attendance. Expulsions, previously unheard of, became more commonplace, as did undercover narcotic cops who were now looking for ecstasy more than for marijuana. One alumnus summarized the impact of social changes in Miller Town during the late 1980s and throughout the 1990s:

> I think perhaps the Miller Town area [was] beginning to realize that it's not isolated anymore, that it has to deal with the outside world, and certainly, seeing the growing pains that go with that.[12]

Miller High students were now taking classes called Education for Responsible Parenting; business classes included introductory courses in marketing; psychology classes had been added; and the popularity of lacrosse had replaced that of football. In the hallways, students kissing passionately were a regular occurrence.[13] Girls showed up to school wearing more revealing clothing than ever before in the history of Miller Comprehensive High School, and students who wore T-shirts advertising alcohol or drugs were called to task.[14]

It was also during this period when the rare young person who openly expressed being gay did so at the risk of great ridicule,[15] and when Miller High students participated in clubs that were religiously and racially defined, such as the Christian Youth Against Drunk Driving, the Christian Young Life Club, the Fellowship of Christian Athletes (FCA), the Jewish Youth Group, and the Black Awareness Club. A bible study group[16] met every morning. At Miller High, at the end of the twentieth century, young people of Judeo-Christian religious traditions openly congregated on school premises to profess their beliefs.

This generation of Miller High graduates inherited the previous generation's youth-defined peer groups, added the punks, goths, and geeks or nerds, and no longer talked about potheads but of drug users and dealers. However, unlike graduates of the Border-Crossing Generation, who perceived themselves and others as belonging to clearly defined peer groups—or to none at all, in which case they identified themselves and others by racial and economic status—graduates of the late 1980s and 1990s remembered themselves and others first and foremost as Christian, Jewish, Russian, black, farm girl, and once again as redneck, superimposing religious, national, ethnic, and class dimensions to youth categories inherited from the previous generation. These categories of identity, however, played out differently for rule-abiding students and those who challenged the system.[17]

Almost all alumni interviewed recalled a very regulated high school setting, depicted administrators as punishing forces, and spontaneously situated themselves vis-à-vis the school principal to underscore their rule-abiding status.[18] Alumni recollections suggested that *not* knowing the principal on a personal level was a gauge of one's consistent good behavior. Knowing the principal, on the other hand, could only mean that one had transgressed. Harry Rice, Sue Cohen, and Cecilia Hood, graduates of 1987, 1995, and 1996, respectively, when asked about their principals, explained:

> Harry Rice: "I cannot remember seeing the principal walk the hallways...was very strict. You stayed away from the office."
>
> Sue Cohen: "I couldn't tell you the name of the principal and vice principal...I couldn't tell you what he looked liked, I wasn't at the office much, I had no reason to be with the principal."
>
> Cecilia Hood: "Never dealt with [the principal], never had to."

Furthermore, some graduates referred to school rules as "the law":

> Mr. L., he laid down the law...If you got caught [using drugs] there was no questions asked, you were automatically expelled and off the school property...on the first day they told you, they set down the guidelines...no drugs, no hitting on teachers...Mr. L., he didn't ask too many questions. If you were caught...you were out the door. Cherry Gate (1988–1992), interview by author, October 23, 2003

Some spontaneously recited a litany of prescribed behaviors:

> There was no drinking, no smoking, no drugs, no weapons, can't bring anything to school that looks like a drug, don't lay a hand on a teacher, you don't hit a teacher, you're not supposed to fight...the basics. Heather Korn (1989–1993), interview by author, February 25, 2003

Of the three generations of alumni interviewed, this one stood alone in its vivid recollections of its school's disciplinary apparatus. Moreover, memories of graduates revealed increasing expulsions across the 1990s and decreasing "school spirit." Memories regarding strong community involvement and pride in being a Miller High student shared by those who attended in the late 1980s transformed for those who attended throughout the 1990s into memories of dim school spirit.[19] Alumna Heather Korn, a 1993 graduate, remembered: "The new kids that went to school behind me, they weren't as involved as we were, they weren't as excited about being there as we were." Testimonies further suggested that administration's hard-line approach to discipline contributed to the lack of spirit. Sue Cohen remembered how the "administration came down really hard":

> There was a tradition, a senior doughnut run, where we run across the street to Dunkin' Donuts...my year, if we decided to do it, we'd get kicked out of

school. The administration came down really hard my senior year...spirit day [when you're supposed to dress alike], nobody did that...nobody came to football games.

Rule-abiding students included those who belonged to religious youth groups; immigrant students seeking social mobility through lawfully sanctioned means; students deeply involved in community service occupations; and usually high-profile students with means, the preps, who continued as they had since before desegregation to craft the school's official representation to the outside world, although, as we shall see, with much less enthusiasm, and much less school-wide participation.

Rule Abiders

Religious Middle-Class Jewish and Christian White Youth

Sue Cohen, a 1995 graduate, belonged to the Jewish Youth Group, and Bill Jackman, a 1999 graduate, belonged to the Christian Young Life Club. Both Sue and Bill, along with friends who shared their beliefs, could be seen recruiting members to their religious organizations on school premises. Sue and Bill recalled:

> Sue Cohen: "We were always trying to recruit people for the youth group. If there was somebody I thought was really nice and Jewish, I would say, why don't you come to this meeting?"[20]
>
> Bill Jackman: "We'd try to get into the school culture...we'd talk to the principal...if you see us around, don't be scared, we're not here to hurt."[21]

Part of their ethos as young people openly committed to their religious beliefs and unapologetic about expressing them within the public school system was an avoidance of conspicuous consumption. Bill belonged to the Fellowship of Christian Athletes (FCA) and was part of a growing number of jocks who refrained from excessive drinking, or from drinking altogether. Within the Young Life Club, he played games once a week, learned some sign language, and listened to or gave "a five- to ten-minute Christian message at the end" of the club meeting. Sue underscored the Jewish Youth Group's adherence to sober behavior when she distanced herself from the stereotype of the Jewish American princess (JAP), which suggests conspicuous consumption and self-indulgence:

> Most of my close friends I had through my youth group... it was the Jewish Youth Group, it was really tight... There were other groups, the JAPs. I hate to say that reference, but they were. I didn't really socialize with them.

These graduates' memories also revealed a keen sense that high school meant little in and of itself, that it was just a "stepping-stone"[22] to get to college, and that it required a purposeful and disciplined engagement.

Sue and Bill attended a combination of honors and some "highly competitive" gifted and talented classes, saw themselves as rule-abiding students, and were fully engaged in extracurricular activities dear to their hearts. Sue invested her energies in journalism, which was "a big deal [to her] at the time," and where she "was the editorial's co-editor."[23] Bill played in the Lacrosse team, which "went to the regional finals a couple of years in a row." Sue also identified herself as "one [of the] nerds," and Bill identified himself as a jock. These were students of solidly middle-class background who understood the importance of investing their energies competitively during their high school years to increase their opportunities for acceptance into the colleges of their choice. Students of the Redivided Generation were very much aware that increasing numbers of high school students, whether they wanted or not, headed for college because it was "harder to get a job without a college degree... they [wouldn't] hire you as soon."[24] Sue and Bill remembered that teachers in honors and advanced placement (AP) classes were demanding,[25] and that competition among peers could be fierce. She recalled:

> The gifted classes were very competitive... A friend of mine was running for valedictorian and someone else was, too. That person took college courses over the summer, had it put toward high school, so that boosted that person's quality point average. The other kid didn't take another advanced course... It was really sad.[26]

These were dedicated students. Sue "cared about doing well and put a lot of stress on [herself]," and Bill "hung out with pretty high achievers." Just as they put the onus of disciplined and responsible behavior upon themselves, whether as young devoutly religious people or as academic achievers, they projected onto all other students the same onus of personal responsibility for academic success. In reflecting on Miller High students and teachers, Bill said that "if kids are willing to learn, they [the teachers] will teach them." Furthermore, for Bill and Sue, high school graduation celebrations such as the prom seemed superfluous. Bill remembered dances as the purview of girls who were usually more involved than boys "in a prom-type thing." Sue remembered them as devoid of any sense of closure: "A lot of people didn't go to the prom, and it didn't matter to us. We just didn't care, it was just school, it was a stepping-stone to get to college."

Their recollections further revealed that being a Jewish or Christian student first and foremost, and secondarily a nerd or jock, considerably narrowed one's world of associations. In an overcrowded school where it had become impossible to know most of your classmates, Sue and Bill, within the spaces of "tight" youth groups and learning alongside "pretty high achievers," lived their high school careers in particular streams of student life that ran parallel to many other streams.

First, they were less likely to intermingle with black students, who throughout the 1990s continued underrepresented in honors and gifted and talented (GT) classes,[27] and they were less likely to intermingle with nonreligious young people. When Sue recalled youth from other backgrounds, they were from another faith: "Christian people came to my bat mitzvah, I went to their confirmations; we all seemed to get along." They were also less likely to encounter sexually active youth or those who did drugs, let alone know them personally. Sue literally did not know whole segments of the student population:

> I don't remember any of my classmates getting pregnant...I heard other people did drugs. I didn't do drugs, so I didn't think [the drug-prevention classes] were useful...I didn't know anybody who had sex.

Bill, on the other hand, made it a point as a religious youth to distance himself from jocks who drank by participating in the Fellowship of Christian Athletes. Similarly, Sue purposely distanced herself from the JAPs—whom she could not completely avoid, since they shared the same secular culture—by actively retreating "into [her] own group" of religious Jewish youth.

While the Jewish Youth Group remained tight and included only young people who actively participated in their Jewish faith, the Christian Young People Club, as Bill explained, included a few Jews and some punks: "We had more than a few punks...and at least three or four kids that were Jewish." Still, the overwhelming majority of participants in the Christian Young People Club included athletes who were also part of the FCA and a lot of preps. Russians, who for the most part were secular Jews, as we shall see, and blacks never joined. In recalling participants in the Young People Club, Bill said:

> A large amount were athletes...and the preppy kids...We wouldn't get cross-racial; I guess there was a separation with the black and white side of issues in this area in general.

While many black youth were fully involved in their faith as Christians,[28] they did not enter the worlds of FCA or Young People Club. In fact, by the end of the 1990s, as Bill remembered, seating arrangements among students in the cafeteria clearly captured streams of students divided along ethnic and racial lines, and revealed parallel worlds that did not intermingle.

> I guess there was a separation with the black and the white side. Usually the black kids would hang out with black kids, the white kids hung out with the white kids, there was some crossover, not overabundance though... it wasn't like everybody that was black hated everybody that was white and vice versa. There were incidents [of racial discord] but not extremes. There were people in school that were racist either way, but it wasn't overall... There were the Asian Indian kids, not black, not white, race-wise, didn't fit anywhere... The Russian kids were definitely a group.

Immigrant Russian Youth

Among the "Russian kids" Bill alluded to was Ivan Strasky, a 2002 graduate. He began his testimony with a description that suggested an essential shift in the historical depiction of immigrant Russians of the turn of twentieth century:

> In the midst of Americans we always spoke Russian. You had to; that was the cool thing to do. Among ourselves, we spoke English.[29]

These were unlike the Russian immigrants of a century ago who struggled to speak "American" English, to present an "American" persona in public, and who in the privacy of their homes forsook teaching the mother tongue to their progeny. Russian immigrants of the 1990s often spoke English before they came to Miller Town, and continued to teach their offspring the mother tongue, fully aware of the advantages of being bilingual in any country in the twenty-first century. Many young Russians at Miller High used their bilingualism as an identity banner to signal their allegiance to "Russianness." Ivan and Vera, Ivan's elder peer who graduated in 1999, were among the rule-abiding Russian youth. These were strategic students for whom high school "was a means to an end." Ivan remembered two very distinct groups of young Russians attending Miller High and how he belonged to the academic group:

> I think there are two distinct groups of Russians. You have the highly academic Russians, immigrant Russians whose parents are always telling their kids, 'School, school, school, we came here for your benefit, we

sacrificed so you can reap the fruits of what we left behind, the only way you can succeed is school.' There are some kids who chose to take their parents' advice and listen to them... [and] others who chose the path of being defensive.

Ivan was an organized, goal-oriented, and highly motivated student who, along with other Russian youth like him who strove to vindicate their parents' sacrifices, single-mindedly pursued college-track courses and built relationships with teachers and guidance counselors along the way.[30] Vera also engaged in studies purposefully and successfully, and although her teacher-student relationships were formally impeccable, she confided that she could not even remember her teachers' names, dramatically suggesting that high school was a quick and swift passage on the way to long and serious studies in higher education.[31]

While Ivan and Vera attended mostly GT classes and immersed themselves in academics, they did not identify themselves as nerds or geeks, as did the white middle-class U.S.-born Harry Rice and Sue Cohen.[32] Ivan and Vera both solidly identified themselves as Russian immigrants. Recollections further suggested that others consistently identified Russian immigrant youth as Russians, not as jocks, nerds, and so on.[33] Thus, whether of nerdish or jockish[34] inclination, a Russian student remained a Russian student. The institutionalized nature of such peer groups as nerds or jocks made those labels U.S.-specific. A freshly immigrated youth remained a foreigner, at least for the duration of his high school years, and as such would not acquire an inherited peer-group identity. Ivan remembered how Russian students huddled together in the cafeteria: "Two tables with Russian kids, each table holds about twenty-five people, a lot of Russian kids!" Vera further recalled how in general it was much more comfortable to remain within her Russian group, where she felt at home, than to risk rejection by stepping into another group. Tired of trying to "fit in" throughout all her middle school, Vera spent her whole high school career alongside her Russian friends and boyfriends:

> All the Russian students hung out together... Actually, until ninth grade, I hardly hung out with Russian kids, I tried to fit, I spoke only English, drove my parents crazy... we spent so much time in middle school trying to fit in, that by the time we got to high school, it was we didn't want to fit in... some of us came when we were twelve or thirteen years old, it was easier to speak Russian... so it was pretty much all Russian people, Russian boyfriends, Russian friends.[35]

Like the GT students Sue and Bill, Ivan and Vera interpreted other students' lack of academic success as a question of personal choice and effort.[36]

However, as Ivan saw it, a particular type of immigrant student was *more* likely to succeed if he applied himself:

> My perspective, it is so much easier to be a Latino immigrant. You don't have to be that hardworking or that intelligent to get here. In the general population that is Mexican immigrants, the really smart ones are the average ones. The Russian and Indian immigrants, people that had to cross the ocean, and had to deal with governments, the iron curtain, it was parents who were witty that found their way through that.

Ivan understood the immigrant story as one of natural selection that privileged, in this case, his national/ethnic group. The initial "wit" required to transplant lives overseas, translated, as he saw it, through "genes and observing their parents" into producing more capable Russian youth. He attributed the fact that one "wouldn't find many Latinos in the upper structure" to the dual role played by genetics and role-modeling. Ivan's use of "overseas hardships" as supporting evidence for his assertion, however, suggested that he was not aware of the oppressive government regimes that many Latinos were fleeing or the life-threatening situations they endured on routes more tortuous than a direct flight from Europe. In his mythologizing of his immigrant story, Ivan did not include that it is precisely the Soviet liberalization of emigration policies of the late 1980s and early 1990s that allowed for the Russian Jew exodus at the end of the twentieth century. As is perhaps often the case for children of immigrants, who may feel obligated to vindicate their parents' sacrifices and who were raised on hearty helpings of hardship stories that tell of all that was endured for their benefit, Ivan constructed his immigrant story as an epic—one he also constructed for Asian Indians, among whom he counted his good friend Prag, who would go on to MIT after graduation.

Ivan proudly recounted that at Miller High, "the achievers are immigrants," Russian immigrants in particular who counted among the "really smart and hardworking Russian kids." While he did remember a couple of Russian athletes, the rest were "that pole of the Russian community [at Miller High] that did nothing extracurricular." It is "that pole" of Russian students who challenged the system and its "American" nature. Whether academically dedicated or anti-American and rebellious, most Russian students dealt with yet another identity layer—that of being Jewish.

Ivan and Vera shared the complex relationship that many Russian students developed around the "Jewish" label when it referred to the religion and not the culture, and they found themselves defending Judaism without knowing much about it:

> Ivan: "If Judaism was ever insulted, Russian kids immediately perked up their ears. The irony in that is that out of the group of Russian kids, none of them that I remember were ever bar mitzvahed. I was culturally Jewish, not by ceremony. Russians in Russia, they were never allowed to practice their religion. They called themselves Jewish; they often didn't have a clue of what the holiday is about... The Judaism has been pulled out of them. They come here, they can practice Judaism, they don't know what or how to do it... The Russian kids are called Jewish because they are Russian, and because they are so loyal [to Russianness], they have to defend Judaism."[37]
>
> Vera: "I didn't fully understand that I was Jewish until I came here, so in my family, we never got into religion."[38]

The label "Russian Jew," doggedly adhered to and flaunted by a segment of Russian students who preferred the mafia world to that of academia, as we shall later see, was often used, as Ivan remembered, like a shield against perceived discrimination by teachers. Ivan explained:

> Whenever the teacher challenged them, not because of being Russian, but because the teacher was being a teacher—"Why didn't you do your homework? Why are you late?"—they would automatically be defensive, and often perceive it as discrimination.

Unlike their defensive peers, and as successful students, Ivan and Vera sported the Russian immigrant label differently, not as a target of ire and reason for defensiveness, but as a vindication of their parents' strife. Vera explained:

> I knew I was going to college for sure, my parents were going to murder me if I didn't go... I graduated with honors. My parents were proud.

Religious beliefs or cultural identification with religion and strategic academic competition grounded the daily behavior of Sue Cohen, a U.S.-born Jew, of Bill Jackman, a U.S.-born Christian, and of Ivan and Vera, immigrant Russian youth who identified as Jews. These high achievers were not the only rule-abiding youth. Many "middle of the road" white students, who were neither members of youth groups nor high achievers, and who did not identify with a particular nationality or ethnic group, but who attended the lower general tracks and more broadly shared a lower socioeconomic status, welcomed zero-tolerance rules that, as they saw it, kept things orderly and safe for the majority.

White "Regular" Students

In describing their principals, Cherry Gate and Heather Korn, graduates of 1992 and 1993, respectively, remembered:

> Cherry Gate: "Mr. L., our principal, he laid down the law and, I mean, you were never upset going through the hallways."[39]
> Heather Korn: "Our principal, she commanded respect...she held her ground and I really liked that."[40]

These were students who took standard courses and whose classroom realities sketched out differently than for those attending the exclusive worlds of honors and GT courses. Bill Jackman, who attended one or two general-level courses and could compare them to his GT and honors classes, explained how "in the standard level, it was a little rougher...The teacher was more afraid, maybe not of students doing something to her, but just from keeping the class from getting too disorganized."[41]

As had been the case for previous generations, it is also within the "standard-level" courses through which the majority of the student body traveled. However, students of the Redivided Generation who lived in the lower academic tracks were more likely to encounter each other less as competitive academic performers and more as representatives of diverse racial, ethnic, cultural, and even class interests and struggles. It is therefore as black or white, Russian Jew or "redneck," subsidized apartment or middle-class suburban home dweller that some of these students competed with each other, and not as college-résumé builders.

It is within the standard-level courses where much of Miller Town's social wounds flared up. It was a place where young people lived closest to the economic realities of their town as work-release students, and as part-time emergency medical technicians (EMTs), firefighters, burger flippers, veterinary technicians, waiters and waitresses, cashiers, tellers, and employees at the many retail stores and corporate giants, from Target to Kmart to Home Depot, that by the 1990s had spread around town. It is also within the standard-level courses that one was more likely to come across drug dealers and witness illegal behavior.

Recollections of graduates of the Redivided Generation revealed growing incidents of violence. However, while those students who attended GT and honors classes barely remembered fights and only heard rumors, as did their counterparts of the Divided Generation, alumni who attended general classes reported feeling the impact of violent incidents more vividly. Heather Korn recalled how "you knew who the drug users were" and that "if you wanted to get drugs, you could have them in ten minutes." She also recalled a day when she was at home sick: "One of the older students brought a weapon to school, a really big gun. We saw it on the news. I was so glad I wasn't at school that day."

To Roberta Jones, Cherry Gate, and Heather Korn, graduates of 1991, 1992, and 1993, respectively, zero-tolerance policies were a welcome protection against what they remembered as increasing school violence. As students attending standard classes, they were more likely than their GT and honors counterparts to witness fights firsthand, in particular racial fights, which increased throughout the 1990s. Cherry Gate remembered, although furtively and regretting having shared her memory almost immediately after uttering it, how her own sister had been the target of racially related death threats:

> A few years after I went to school, yeah, my sister was supposed to graduate back in '96, and she actually got a death threat and she went to Springfield High. She moved because she was scared to go and stay at the same school because at that point in time it was the racial fights going on.[42]

While Cherry, who graduated in the early 1990s, suggested animosity from black youth toward her sister, Bill Jackman, who graduated at the end of the 1990s, spoke matter-of-factly about white as well as black racism, suggesting that as the 1990s progressed and racial and ethnic tensions escalated, many Miller High white youth considered "racism" no longer singly the sin of the whites.[43]

Still, Bill, who attended during the second half of the 1990s, the years that Cherry suggested were marked by "racial fights going on," and who attended mostly honors and GT classes, recalled racial fights as being incidental and far from representing the norm. The disparity in Cherry's and Bill's assessments of the severity of racial problems at Miller High could be interpreted, as earlier discussed, in terms of Bill's attending the removed spaces of GT classes, which made him less likely to witness violence among peers than if he attended standard courses. However, there is also reason to believe that Cherry's sister's involvement in the racial incident she recounted might have colored her interpretation of the extent and nature of racial problems at Miller High. While her testimony implied that the onus of the racial problems in her sister's case lay squarely on the threats from black students, it wasn't until almost a year later when I interviewed Cecilia Hood, who graduated the year Cherry's sister would have had she not gone to Springfield instead, that I reframed Cherry's testimony in terms of possible racist animosity by Cherry and her kin against newly arrived blacks. Cecilia recalled:

> I think twice in the course of my high school there was a big racial fight between the black kids and the farmers... Two kids from Miller High were sent to Springfield High, got expelled, and two others got expelled and were not allowed back.[44]

Cecilia's identification of the whites involved in the racial fight as being "farmers" fit Cherry's self-description as "a back-home-farm-country girl" who "worked at [her] parents' farm."[45] Cecilia's memory of two kids being sent to Springfield High matched Cherry's mention of her sister going to Springfield High. Their accounts differed only in that Cecilia remembered that the students had been expelled and sent to Springfield High, while Cherry's recollection implied that her sister had chosen to go to Springfield High because the racial tensions at Miller High had become untenable. Finally, Cherry's refusal to elaborate on the incident or to provide access to her sister further cast suspicions on her sister's total innocence in the racial incident. Then, too, it's quite possible that Cherry did not want to provide contact information for her sister simply because her sister harbored such bad memories regarding her transfer (or expulsion). Confusion about Cherry's sister's involvement in racial disputes notwithstanding, Cherry's and Cecilia's conflicting testimonies illustrate the escalating reality of racial tensions at Miller High during the 1990s between "farmers," as they called themselves, or "rednecks," as others pejoratively called them, and blacks as well as Russians.

Beyond their shared feelings of security under principals who "laid down the law," Cherry, Heather, and Roberta placed themselves somewhere between "the group that did drugs" and the "higher group... that you knew were really smart."[46] These alumnae remembered themselves as unassuming, down-to-earth students who hung out with the "standard" group.[47] These girls distanced themselves from the preps and those with money and constructed the distancing as a matter of moral choice. Cherry recalled:

> I never hung... I guess with the popular group because I didn't believe in their status, the way of doing, the way of thinking. It was about the right clothes, the right makeup, the right hairdos... they were up on the *Vogue* and stuff, you could feel they were snobs.[48]

While memories of Divided Generation alumnae highlighted the importance of "looks," and while memories of the Border-Crossing Generation alumnae played down the importance attributed to "looks" in general, the memories of these 1990s alumnae stressed the emotional tensions that young women experienced around issues of clothing as they reflected on one's economic status. The identity of the "prep" had been synonymous with upper-class status and high visibility in all social affairs pertaining to school across generations. However, recollections of alumnae of previous generations did not in general linger on the preps. One alumna, Sophie Baker, who graduated in 1985, at the very end of the Border-Crossing Generation, did remember the preps negatively, as earlier discussed. By the late 1980s and 1990s, the prep was not only openly criticized, but also avoided.

In Cherry's and Heather's eyes and those of their friends, the preps were vain people, exclusive rather than inclusive. By contrast to preps' undemocratic attitudes, Cherry and Heather saw their ways as solidly down to earth. They knew themselves as "regular people" who considered everybody as equal. Cherry recalled:

> My friend was also the same. She was middle class...she stayed away from them [the preps], she felt the same thing I did, and the groups that were just basic regular people. You could feel that they [the preps] were snobs. In other words, they would make you feel that. Okay?

Heather also recalled: "They had the snobby attitude, but they were no different than you and I."

There was yet another group of students that Heather in particular found offensive—to the senses: the Russians. Echoing the late nineteenth and early twentieth century "American" complaint against many immigrants' "poor hygiene," Heather recalled: "We had a large group of people who were not clean people, like they didn't shower."[49]

For these young women, who attended mostly general courses and worked in the community after school,[50] who felt shunned by the preps and unsafe without rules to protect them against students apt to engage in illegal drug dealing and violent behavior, high school graduation *was* the end of their student career, at least for the immediate future following it.

Cherry, Heather, and Roberta constructed their high school days not as "a means to an end" or a "stepping-stone to get to college," as their peers in GT classes might have, but as a process of maturation with graduation itself marking the end of the life stage of adolescence. Their recollections suggested young people who had fully integrated a sense of themselves as incomplete humans on the way to maturity, and had associated the idea of high school graduation as a marker of their new identities. While Roberta Jones lamented having to grow up, Cherry

Gate celebrated the process:

> Roberta Jone: "It's really hard being a teenager, because you're in the middle... still being parented by your parents, and it's hard now [at graduation] you know you got to do something."
>
> Cherry Gate: "I grew out of my shell into a better person."

These alumnae's recollections echoed to a certain degree those of the Divided Generation who attended the general or commercial tracks and who saw the end of high school as a rite of passage and the high school diploma as a legal tender. At graduation, one had the right to assume one's own independent life without stigma. Heather and Roberta did go on to community college; however, Heather, right after graduation, and Roberta shortly after, found that their high school diploma limited their job options and, along with Cherry, remembered experiencing the end of high school as a place where one left as an adult ready to assume adult roles.[51] The "standard classes," however, as Roberta's recollections suggested, were also the place to which young people went who didn't like school and who didn't know what their lives were yet about.

> I let everything interfere with my schoolwork. I didn't like it...I wasted all this time on something I didn't really enjoy...I just don't do well in school[52]

For them in particular, school was a no-win situation. Neither staying in it nor leaving it brought any sense of relief. Leaving it meant that "now you know you got to do something," yet you might still not have known what you were supposed to do or be outside school walls. Roberta recalled: "It [high school] still didn't help me pick out what I wanted to do." Neither ready to assume a place in the job market nor headed for college, and feeling as if she had wasted her time within high school, Roberta could not find a socially defined acceptable space within which to place herself. At graduation she was an adolescent required to be an adult in a world for which she felt unprepared.

Having spent their high school years closer to the job market than their GT peers, closer to girls who got pregnant and boys who enlisted, closer to drug dealers and social prejudice, it is little wonder that these students would have understood the end of high school more in terms of entrance into adult life than their GT peers, whose dependence on parents would have to be prolonged as they pursued undergraduate and graduate studies. For the standard-class rule-abiding students, high school still held some sense of finality. Many of these alumnae's male friends were part-time firefighters, emergency medical technicians, and sometimes work-release students,[53] young people whose lives daily and weekly were involved with the community. Inevitably they would soak in much more directly the effects of economic competition and social pressures, and integrate, perhaps even perpetuate, the social prejudices of their surroundings.[54]

The Preps

By contrast, preps' experiences of Miller High in the late 1980s and throughout the 1990s were dramatically different. Preps continued to be identified by their economic status and the high profile they assumed through their pervasive involvement in the school's extracurricular and social life. In describing the preps, Bill Jackman accentuated their wealth and isolated upbringing:

> Kids that knew each other their whole life, they grew up together, a lot of those kids played sports, that connection... they do have a decent amount of money... they did the same things growing up and people don't like to change... a lot of them dressed preppy, I think it was more they were used to hanging out with each other.

Russian immigrant Vera Debin also accentuated their wealth:

> Their parents were wealthy, they had the best clothes, the trendiest clothes... the wealthy kids hung out with wealthy kids... the preps.

Cherry Gate stressed their upper-class status and visibility, and underscored, as Bill had, their tight bonding:

> The popular group, I guess we would consider like the upper, upper class... I would say their appearances and everything would make them bond together.

Recollections of two alumnae, Cecilia Hood and Betty Ames, graduates of 1996 and 1999, respectively, suggested that during their high school years, they had been the preps that Bill, Vera, Cherry, and others talked about. While neither Cecilia nor Betty identified themselves directly as preps, although Cecilia identified herself as being part of the popular group, both alumnae were of upper-middle-class background; fondly remembered their high school years as a time when they were involved in myriad sports, extracurricular activities, and leadership roles; and reported enjoying long-term friendships with peers whom they had known since early childhood. Betty explained:

> I played two sports every year, I was class treasurer, I did meetings for that, I was also in SADD [Students Against Drunk Driving]...I was in GT and AP honors classes...I was in school every day from eight in the morning and I didn't get home until 5:30...The class officers, we put a lot of effort into organizing [school events].[55]

While Betty and Cecilia held the same high profiles as their counterparts of earlier generations, they did not enjoy the same levels of participation in their organized social events. The worlds created by the preps, which had been patronized by the Divided Generation and tolerated by the Border-Crossing generation, were being boycotted by many non-preps of the Redivided Generation.

Over the three time periods studied here, there is evidence to suggest that school social events, prom in particular, had been the playground for Miller High's white female preps. Proms were a symbolic space within which, across time, the preps exerted progressively greater control over all aspects, even to the point of rigging the results of school-wide elections for the prom song to impose their own sentiments against majority vote, as Sophie Baker remembered about her graduating prom of 1985. The preps continued also to represent the school through yearbooks, without necessarily representing all students. Vera Debin explained how biased the yearbooks they created were:

> I think they [the preps] went around and tried to get pictures of all the clubs, and all the band, but it was pretty much them and their friends, all extra pictures.

In her testimony, Vera suggested that yearbooks would "probably be more interesting if people were assigned to a committee to do the yearbook, randomly selected. Across the three generations of Miller High, recollections suggested that preps image-managed their school.

However, while Linda Moss, a 1969 graduate, who remembered herself as one of "the doers...the people that were on the teams...putting out the yearbook...class officers," enjoyed a "big school spirit" and great participation in all school events that she helped organize, Betty Ames, thirty years later, felt frustrated by the lack of student participation. She remembered:

> The class officers, we put a lot of effort into organizing [school events.] It'd be frustrating when people didn't participate....School spirit was terrible.

Preps continued to organize the visible social life of Miller High, but over time, fewer and fewer people showed up, and there is reason to believe that the prom became the event that non-preps could boycott, symbolically rejecting through nonattendance the world of the prep, the one peer identity primarily defined by "upper-class" socioeconomic status across generations.

When Sue Cohen remembered not caring about going to the prom, she also might have been, as a religious youth who distanced herself from excessive consumption, rejecting the ostentatious nature of proms. Over time, as yearbook pictures revealed, proms had become progressively more elaborate productions. By the 1990s at Miller High, they had evolved into showy productions that involved chartered limousines and expensive attire.[56] Russian immigrant alumna Vera Debin was among those who barely attended her prom. She peeked into the ballroom and almost immediately left with her boyfriend. She felt that it was merely a congregation of the prom organizers to show off their expensive clothes. She recalled:

> I wanted to see with my own eyes, walked in, saw what it was about, took pictures and left.
> *What was it about?*
> The [prep] cliques again, everybody looking at each other's dresses.

Preps' growing monopoly over time over the prom phenomenon might have contributed to institutionalizing the prom as a prep phenomenon. The institutionalized nature of the dance, inherited from the turn of the twentieth century,[57] would have been difficult to ignore. Yet by the 1990s, youth with strong religious convictions, some Russian immigrants and some "farm-country girl[s]" might have found it difficult to participate in a primarily prep-devised extravaganza, a ball for the princesses where king and queen are crowned. Furthermore, the general lack of school spirit throughout the 1990s and the many student nuclei, where "people kind of stayed in their groups"[58] and seldom crossed over between groups, might have made school organized parties by the few less appealing.

Although by the 1990s, as recollections of non-preps as well as preps suggested, the preps had become an elite group more unto themselves than at any other time since the Divided Generation, they were no longer only white. Black and white preps lived parallel lives, and occasionally intermingled. Bill Jackman and Vera Debin explained:

> Bill Jackman: "There were two popular groups. The popular black kids and the popular white kids. They didn't look down on each other, they would still interact and stuff. There was definitely two sides."
>
> Vera Debin: "A lot of the popular kids among the black kids would be friends with the [white] preps."

African American young women continued to fill the ranks of cheerleaders since the Border-Crossing Generation, and by the 1990s were also homecoming queens.[59]

While race, nationality, culture, and academic tracks primarily defined student relationships for blacks, whites, and Russians of the Redivided Generation, a segment of students across race, ethnicity, and culture congregated as youth of means who held high profiles. Thus, while African American youth of means Nat Right, a 1960s graduate, would be perceived by his white peers in the general track as "poor," by the 1990s, economically well-to-do African American youth were perceived as such by students across tracks. The prep was an identity primarily wrought around one's socioeconomic status. For preps, class trumped race and ethnicity. Russian immigrant Vera Debin recalled one Russian peer "who came here when he was a baby, his parents were very well off, he pretty much hung out with them [the preps], he hardly spoke Russian." As a Russian then, the less immigrant one was, the more money one had, the easier one became a prep. Under the umbrella of upper-middle-class status, the categories of race, nationality, ethnicity, and religion merged at Miller High. If one were wealthy enough, one could be a black, Russian, Jewish, or Christian prep, and more easily rule abiding, since one's personal life would be deeply interwoven with myriad school-sanctioned activities that one would participate in developing and nurturing.

When asked to reflect on their peers in general, preps Betty and Cecilia recalled, as had other graduates of the Redivided Generation, how students segregated into groups:

> Betty: "Everyone definitely segregated themselves...in the cafeteria...two tables of black kids, and then the Russian kids, the athletes...definitely students grouped within their cultural backgrounds."
>
> Cecilia: "People kind of stayed in their groups...The black kids did sit together. We started having a large population of Russian kids."

They also recalled accounts of racial fights, which they, unlike their peers in the general track, never saw close up. Commenting on an article that she had read in the *Community Times* about racial fights at Miller High, Betty remembered: "I never noticed [racial tension]. It could be that I was removed from anyone involved in it." Cecilia, as earlier discussed, remembered the expulsions of students who were transferred to another high school because of their involvement in racial fights, but she had never witnessed them firsthand. Preps Betty and Cecilia lived their high school years alongside incidents of violence that did not directly affect them. Rule-abiding students, they engaged in leadership positions and sports, and lived lives parallel to those whom some alumni remembered as "troublemakers."[60]

Just as students of lesser means remembered the preps and how they dressed, Cecilia and Betty also remembered students with lesser means, and how *they* dressed:

> Cecilia: "There were a couple of kids that didn't have a lot of money. They wouldn't dress so great."
> Betty: "Boys that were involved in the fire department, they wore flannel shirts and jeans—country boys."

While the 1970s generation paid little attention to clothes, in the 1990s, just as in the 1950s and 1960s, clothes and cars played captivating roles in the imagination of many students. Speaking from the prep's point of view, Betty recalled: "A lot of that behavior [people making fun of people] came from the cars people drove. People turned fifteen and their parents would get a brand-new Explorer."

Betty and Cecilia, just like the "upper class" and "doers" of the Divided Generation and the preps of the Border-Crossing Generation, "always expected to go to college," whether they invested in the grades or not.[61] Betty's recollections in particular revealed her loyalty to Miller High, a school that, like the preps before her, she represented to the world and to the school's students through yearbooks; a school in which she invested many extracurricular hours on social events to which too few students came. Speaking of her alma mater, Betty said:

> I can't imagine my life without it...I definitely felt that Miller High was a great school...I felt like I got a good education, probably better than at a private school.

While remembering the preps first and foremost as wealthy and self-centered, Russian alumna Vera Debin, with a sense of concern for fairness, said:

> They are the people with the school spirit. I can see why they are representing the school.

Black Youth

Although some wealthier African Americans counted among the preps and intermingled as preps, most African American students segregated as black students. David Randle, in commenting about the 1990s in Miller Town, drew a distinct line between African Americans who had grown up in the town and those who hadn't, and between his generation and the new generation of young blacks impatient with the white status quo.

> You've got to look at our situation. We grew up in the county. We kind of knew how things worked. We kind of knew what to do and what not to do. And then you've got twice as many people. I mean, what happened to the black population at Miller then [in the 1990s]? It more than doubled. So you're talking about a whole lot of people. I mean they came from the outskirts, they came from the city, and they didn't want things to change, or they wanted things to change faster.

When I asked David if I could interview his daughter, who was finishing up high school, to get an African American high school student's perspective of the latter part of the 1990s, he said that she would not want to be interviewed by me. The same occurred when I asked another African American Miller Town resident if I could interview her much-younger brother, who had attended Miller High in the 1990s. The implications of my whiteness within a U.S. reality, in conjunction with my PhD candidacy in education at the time of the research and the power differential inherent in the production of research interviews created in the eyes of two potential African American narrators, both graduates of the 1990s, a person to whom they would not want to grant an interview. They explained to me that their daughter and brother, who are "militant" about race issues, would not want to be interviewed by a white researcher. The agency of these two alumni cannot be ignored. They refused to give their insights and interpretations to someone whom they understood to represent an educational establishment that has failed black students. Their political statement reconstructed the power differential on their behalf. Their refusal to tell of their experiences is a historic act that must be recorded as clearly as remembered testimonies. It could have been omitted or, perhaps worse in this case, footnoted. Their refusal is part of the metadata that deepen the history of Miller High student relationships, but inconveniently interrupt the narrative.[62]

David's daughter's generation approached relationships with whites suspiciously, skeptical about the realization of racial equality in the United States. Recalling his experience as a coach at Crescent High, a county high school predominantly attended by black students, David recalled when their football team played Miller High in the late 1990s. His recollections revealed a racist principal.

> We had a game at Miller High. I'll tell you the name, name is Thomas Lawrence, the principal. I'm standing on the sideline with the principal of

Miller High. Crescent High is playing, and there's a penalty, and the first thing he says is. "Oh, here *they* go again" [David's emphasis]. I mean, how am I supposed to feel... I knew exactly what he meant: "They're all black so something is going to happen; they're getting ready to start something."

By contrast, the recollections of white "standard student" Heather Korn, a 1993 graduate, suggested the opposite—an administration that privileged black students to the exclusion of white students.

One day they had a speaker come. All the black students were invited to go to listen to the speaker. None of the white students were allowed to go. They [the black students] were taken out of class. It was a big deal because, at the time, they [the black kids] were having a big fit, [it was] not too far from the Rodney King beatings. So a lot of people started saying if we only had something for the girls, or only for the whites, there would be a big problem. We're just sitting back and watching this happen.

Heather's either purposeful or slip-of-the-tongue use of "they were having a fit" as she referred to black students' feelings about the broadly televised cruel beating of Rodney King betrayed racial animosity on her part. Heather's statement, "If we only had something for the girls, or only for the whites, there would be a big problem. We're just sitting back and watching this happen," captured both the late twentieth century constructed concept of reverse racial discrimination, as well as the extent to which racism lived close to the surface in the 1990s at Miller High, no longer policed or buried deep as it had been during the Border-Crossing Generation. Since she suggested reverse racial discrimination, perhaps Heather did not know or ignored the fact that the Bible Study Group, which met daily, was attended only by white students who represented, as Vera described, "the ultra-religious right."[63]

Interracial Athletes

While black students mostly sat together, it was on the sports fields that students worked as teams and forged friendships across racial and ethnic backgrounds, as it had been since the early days of desegregation. Betty Ames recalled, "[A] lot of the basketball players were black, and the white basketball players would hang out with them." Still, it was only within certain sports that black and white students met in the 1990s. Throughout the 1990s the football team fared poorly,[64] and more attention went to basketball, soccer, and lacrosse. Black male students played neither soccer nor lacrosse,[65] so basketball was the only sport other than football where they played alongside whites. Throughout the 1990s, football lost the prestige it had enjoyed during the Border-Crossing Generation. Yearbooks, however, revealed black female students on soccer teams, as well as on softball and field hockey teams. Cecilia, who played soccer and lacrosse, remembered that her group of female jocks included "a group of black friends and two friends who were Indian."

White Youth Subcultures in the Midst of Racial, Ethnic, National, Class, and Religious Segregations

Peer groups of the Redivided Generation, unlike those of the Border-Crossing Generation, were white, and were local manifestations of broader international youth subcultures. Indirectly, through some graduates' memories, the punks and goths came into view. Vera Debin remembered the punks as defiantly anti-prep (the one group primarily identified by its superior buying power and upper-class status): "A punk would not go and talk to a prep, because he doesn't need their acceptance." In contrast to Vera's recollection, Bill Jackman remembered how one punk in particular crossed group boundaries:

> He was a punk kid; he was in a punk band, traveled around the country. He had a Mohawk. He was the most extreme-looking kid out of the whole crowd. He crossed over in all the groups. Every teacher loved him. He was a great kid. You look at him, you run away from him.

These were suburban youth whose affiliations with the punk subculture might have been more aesthetic expression than genuine economic and political protest against forms of commodity.[66] Still, Vera's recollection that a punk did not need the preps' acceptance underscores a more broadly shared reaction to consumerism and conspicuous consumption by a growing segment of Miller High students. In this sense, Vera's remembered punk shared with Sue Cohen of the Jewish Youth Group, and with Bill Jackman of the Fellowship of Christian Athletes and Christian Young People Club, an aversion of and challenge to consumerism. Of interest too is that some punks visited the Christian Young People Club, as Bill remembered: "We had more than a few punks [attend the Christian Young People Club]." Because the Christian Young People Club tended to be frequented by youth of politically conservative backgrounds, and in view of the one punk whom everyone loved, it seems safe to assume that Miller High punks espoused an aesthetic rather than political expression of the broader punk subculture. Thompson describes the suburban punk as follows:

> At the end of the continuum is the suburban youth who works for his spending money at the Hot Topic store in a mall, wears the store's apparel, and sells major label–produced CDs by Blink 182 and the White Stripes. This youth's cohort includes everyone who encounters punk primarily through videos on MTV or VH1. At the other end of the spectrum are the 'genuine' punks...who listen solely to independently produced anarcho-punk, steal or panhandle for their means of subsistence...and eschew most forms of commodity.[67]

Along with the punks, goths, who are generally nonpolitical,[68] constituted youth subcultures that were white, middle-class, and suburban. These were consumers of already commoditized pop cultures, even as some of the punks rejected the prep status. One seriously wonders whether school authorities or peers would have as easily accepted an Afro-punk.[69]

Whether rule-abiding punks, preps, or immigrant Russian Jews, or whether "down-home farm girls" or religious youth, these young people lived lives that ran parallel to those known as "troublemakers." Troublemakers, as Betty and Vera described, were targeted by the administration:

> Betty: "They [the administration] knew who the good kids were and the kids that got into trouble. They [the administration] were looking for the kids that got into trouble all the time."
>
> Vera: "The troublemakers would get treated differently."

Troublemakers

Some rule-abiding Miller High graduates of the end of the twentieth century identified troublemakers in generational terms. Roberta Jones, who graduated in 1991, recalled:

> I just remember the last couple of years in high school the younger kids seemed to be a little bit more out of control...I didn't think they were as eager to be friendly with teachers, they were more trying to get into trouble...I think that by the time I was a senior, I think the freshmen coming in, it just seemed like they didn't care as much. Yeah, back-talking—"I don't have to do it,' 'I don't care if you send me to the principal."

Drug users and drug dealers featured prominently as rule breakers in the memories of rule abiders of the Redivided Generation. Under zero-tolerance policies and organized efforts to create "drug-free zones," punishment for these rule breakers was swift. Cherry Gate recalled: "We always made sure it was a drug-free zone. If you got caught you were expelled and off the school property."

As the decade moved along, troublemakers took on more distinct identities and emerged overwhelmingly as Russians, blacks, and those referred to as "rednecks." Troublemakers were "standard-class" attendees. Reporting on Russian and black animosities, Ivan Strasky explained:

> Russian people were not tolerant of African Americans. Mostly Russians here are Jews, who came over because of discrimination. You'd think they'd be open-minded... endless cycle of discrimination... the most clashes came between the Russian community and African American.

Still, as earlier mentioned, recollections also indicated that clashes occurred between black students and the farmers. But some farmers, called "rednecks," also fought with Russian students and egged them on by saying, "Russian, go home." Ivan remembered:

> What has happened is at Miller High there is a distinct community of students who would be called, quote unquote, "rednecks." That is just what people generally say, people who are close-minded, hate Jews, hate blacks, could be categorized as KKK types. So that subpopulation at Miller High would threaten, by just being there, the Russians, because of their Judaism, and because of their being Russian. It's all kind of a broad assault on Russians.... On occasion I have heard slurs about Jews directed at Russian students.

There is reason to believe also that while clashes between "rednecks" and blacks, and "rednecks" and Russians, occurred primarily because of racist attitudes, they might also have occurred for economic reasons around drug dealing. Some graduates' recollections suggested that in the perception of Miller Town old-timers, the arrival of city blacks increased drug-overdosing incidents among black and white students as well as drug deals perpetrated partially by black "newcomers." Drug dealing, however, as Ivan's testimony suggested, was also taken up by many Russian youth who fancied themselves "Russian Mafiosos," wore "their leather jacket, stern look, a cigarette, and a defensive approach to things." Ivan recalled:

> I have heard stories about Mafia this and Mafia that. Kids pick up on this idea of Russian Mafioso... Russian kids, they are smart, Unfortunately their business was drugs.

It is reasonable to assume that Russian and African American drug dealers might have perceived each other as enemies when competing for the same market, especially in view of their loyalties to their respective racial and ethnic groups. Ivan said, "There is a very strong cohesion in the Russian community... Russian culture is very loyal."

Beyond cultural loyalties, many of the Russian troublemakers and "troublemaker black kids" shared two other characteristics that might have contributed, ironically, to their animosity toward each other. Students of both groups tended to come from the city:

> A lot of the Russians moved from the cities to suburbia, which is a big shock. In the city, it is a tight community. You walk everywhere, go to café, bar, everything was city oriented, Then, hit suburbia and oh, what do we do![70]

This youth had been transported from the high-energy life of the city to the comparatively lackluster life of suburbia. While Miller Town was suburbia in the midst of accelerated urbanization, it lacked sites of action that a city proper offered: "What do we do!" One can imagine that these young people itched to make something happen in their immediate surroundings. Proud,[71] loyal, and restless, they also shared a fundamental mistrust of the American educational system. Russian youth had a "lack of confidence, lack of faith in the American system." Ivan explained:

> They were that pole of the Russian community. Nothing extracurricular; they were not committed to going to college, or if they were, they had a negative perception of school, that they didn't want to have anything to do with school...
>
> There is a strong anti-American feeling... what can an American teacher authority teach me?

Russian "troublemakers" could often be seen arguing with their teachers, and teachers would anticipate their challenges:

> From the way I know teachers and the way I talked to them after the years I left, there is this perception that teachers definitely know the difference between a Russian kid and average American Joe. I don't think teachers ever treated kids differently willingly... but when students started arguing with the teacher, the teacher would be more likely to butt heads with the Russians... just because it was almost expected, because us Russian kids would always argue pretty well, I don't know if that is Judaism or the Russian culture... as soon as the teacher sees she is losing an argument, she uses her authority. There is a predisposition against Russian kids, because that general stereotype.[72]

Similarly, some African American students of the Redivided Generation might have felt that black people had been overstudied by representatives of the white race, who in turn had too little to show for their research in terms of advances in the education of black youth in the United States.[73] Pride and loyalty to their respective racial/ethnic cultures, city restlessness, and fundamental mistrust of the American educational system could have made the black and Russian troublemaker students allies in rebelling against "the system," but it didn't. Racism on the part of some of the Russian youth, and perhaps competition in the drug market, made rivals out of these rule breakers. Furthermore, young Russian rule breakers seemed to revel in competing to break rules first, and to have things "happen sooner to [them]":

> First to get drunk, the first to [smoke] cigarettes, the first to use marijuana, the first to use ecstasy...huge use and abuse of ecstasy, a lot of people fell into that, that always happened earlier in the Russian community.[74]

There was also the question of defending one's identity:

> I don't think the African American community ever started anything. It was defensive, on the African-American part, just as it was on the Russian part, just as it would be in any minority. Defend their identity.[75]

Male rivalries had dominated the memories of previous generations of graduates, but by the 1990s, recollections emerged of girls fighting.[76] Black and white girls fought their own racial battles. Heather Korn recalled:

> It was a black girl and a little white girl who was very snotty, she made it known that she was very racist. She was in the cafeteria line, she looked at this girl funny, and the black girl got ticked off because the white girl looked at her funny. Somewhere in the middle of everything, while we were eating lunch, you heard somebody scream, then you heard loud bangs. The [black] girl came up behind her, grabbed her hair, and slammed her face into the cafeteria table. They both ended up getting suspended, so it wouldn't look like a racial issue.

Girls could be seen "scratching"[77] while fighting over boys. Roberta described: "There was one fight that was outside of the lunchroom...It was two girls, probably fighting over a boy." Ivan described Russian girls who might count among the troublemakers and rule breakers as: "Strong Russian bitch. Don't mess with that girl, she will talk you off." By the end of the 1990s, males and females equally engaged in physical fights and verbal aggression.

Exacerbated Segregation

By the end of the twentieth century, Miller High students segregated not only along racial, class, religious, and ethnic/national categories of identity, but also across those categories, as rule abiders and rule breakers. To complicate the matrix of segregation, students were further divided in their experiences of each other and school authorities along "standard classes" and "upper tracks," as they had been in the 1950s and 1960s. In the following chapter I discuss the self-segregating tendencies of students of the Redivided Generation in light of structural forces, and I examine more closely the clashes in student ideology between youth who espouse consumerism and those who oppose it while remaining overall a politically inactive student body.

Chapter 6

Oppositional Self-Segregation: A Student Body Sensitized to Discrimination (1986–2000)

For the Redivided Generation, class and race—divides to which this generation also added those of religion, ethnicity, and nationality—were no longer categories of identity by default, as they had been for the Border-Crossing Generation, but rather public statements of allegiance to particular social interests and world views. This was a purposeful generation not so much interested in "getting along" across categories of social identity as was the Border-Crossing Generation, but more of asserting differences across social categories. Being African American, a Russian immigrant, Christian, or Jew; working class, middle class, or preppy and rich were identities embraced proactively and even defiantly in opposition to other identities. Moreover, the particular geography and demographics of Miller Town by the end of the twentieth century, now an urban suburban town with leftover rural-suburban roots, further divided residents into old-timers and newcomers, divided African American residents along generational lines, and divided whites along national lines.

These divisions along race, class, nationality, ethnicity, religion, and generation, as discussed in the previous chapter, were not uniformly clear-cut. That is, significant wealth trumped racial and ethnic divisions, since well-to-do black, white, and Russian youth socialized as preps. As had been the case for students of the Divided and Border-Crossing generations, athletes crossed racial lines on sports fields, as did students attending the prestigious upper academic tracks. Preps, athletes, and upper-track students notwithstanding, many more students of the Redivided Generation segregated within standard classes, where they also engaged in racial fights. This was a generation that adopted the broader social categories that define census data for new purposes—for personal defense and empowerment, albeit not broader political change.

Peer Relationships

Balkanized Student Groups

Miller High students' self-segregation of the late 1980s and 1990s resurrected stereotyping, something that had been fought against by the Border-Crossing Generation and suffered by the Divided Generation. Russian immigrant Vera Debin recalled:

> I remember some incident, somebody got into a fight, with one of the [Russian] guys in our group, and then his mother got a restraining order against the Russian boy because she was afraid that the rest of the Russian kids would come to her house and beat her boy up.[1]

Russian youth were more likely to be stereotyped as aggressive or even criminal; black youth as drug users; white farm youth as rednecks; and religious youth groups, who were overwhelmingly white, as ultra-right-wing conservative. But these stereotypes were renegotiated by the students in ways that defined their loyalties to particular group identities within Miller High and the broader community, generating myriad combinations of oppositions. Recollections, as discussed earlier, revealed some of these identifications in contrast: a native-born Jewish religious youth defined in contrast to the stereotype of the Jewish American Princess; a Christian religious jock defined in contrast to overindulgent, alcohol-drinking jocks; a Russian immigrant youth defined in contrast to an "American" youth; a troublemaker immigrant Russian youth defined in contrast to an African American youth; a white farm youth defined in contrast to black, Jewish, and Russian youth; and segments of all youth of lesser means in contrast to the preps.

The particular demographics that both significantly increased and diversified the population of Miller Town and its high school, the stringent disciplinary technologies delivered by distant school authorities, the ever-escalating tracking,[2] the broader intensification of the credential race and the undemocratic power of the preps all combined to create an environment conducive to the balkanization of student groups and to students' disengagement from school life.

Explaining the Balkanization: The Intersecting Roles of Demographics and School Structure

The sudden and overwhelming influx of ex-city dwellers and immigrants to Miller High drastically increased the number of students who were not only new to the neighborhood, but who also had not practiced peer relations within desegregated schools, because they came from myriad schools beyond Miller High feeder schools—from the city and from Russian schools. Youth of the Redivided Generation were doubly strangers: to each other and to neighborhood kids attending Miller High. The school was also doubly strange to them: directly, because they did not grow up attending the town's fall festivals organized on its premises or the games and presentations in its gym and auditorium, and indirectly, because they did not have older siblings and relatives to relay school traditions or gossip about teachers. Not surprising, as newly arrived outsiders within an overcrowded Miller High student population, these young people sought the company of students with similar backgrounds. As Russian immigrant graduates put it:

"It was easier to speak Russian."[3]
"I was drawn to the Russian crowd...hanging out with Russian friends, dressing the Russian way...speaking Russian."[4]

Alumni testimonies across time overwhelmingly revealed that when entering Miller High without prior or subsequent involvement in extracurricular activities, students of similar racial, class, and ethnic backgrounds naturally tended to congregate.[5] This was particularly the case for Miller High students in the late 1980s and 1990s. Amid the overcrowding, they huddled in groups for comfort: the Russian immigrant students for whom "America" was still foreign, and the religious students for whom mainstream U.S. culture was too lax in its mores. Other students huddled together out of comfort: the preps who simply knew each other forever and traveled the same upper-class circles; the African American students, newcomers and new generations, who no longer adjusted their attitudes to fit the white world of Miller Town as might have the previous generation, who "knew what to do and what not to do"; and the "farmers" or "rednecks" who felt besieged by ex-Communists, Jews, and blacks all at once and wanted people to "go home," as recollections of others impressionistically sketched out.

Graduates across time periods consistently reported that students associated with certain groups and not others out of a need for shared comfort, shared upbringings, shared language and musical tastes—that is, they congregated for cultural reasons. In the 1990s in particular, in the context of a seriously overcrowded school and an institutional environment that exposed its students by stripping them of privacy to the point where students' personal effects could be searched if administrators found them suspect, finding comfort was difficult. Comfort became the other body whose language and cultural traditions reminded one of home. Youth group associations wrought around familiarity were the most natural and spontaneous expression of the human need for shelter and comfort in the face of institutional discomfort created by overcrowding, intolerant and strict punitive disciplinary strategies, and an influx of foreigners.

However, for a segment of African American and white youth of the Divided and Border-Crossing generations, the comfort of shared culture had always been inextricably linked to race. While this fusion of race and culture was not set in stone for the earlier generations, it became more rigid for the Redivided Generation. Being a black Miller High student definitely meant being of a different culture than a white Miller High student. Many young county blacks at the end of the century submerged their own particular upbringings under the racial category of "black," and identified with ex-city black youth—often to the dismay of the earlier generation of long-time black county residents, who lamented losing their young to drugs and the bad influences they attributed to city blacks. One African American mother, a long-time county resident, said, "I don't want them blacks from the city coming into my town ruining Miller Town. That's how old blacks in this area feel." Many white students stereotyped as "rednecks" also conflated race and culture: Being white American meant being anti-black, anti-Russian, and anti-Jewish. For many white Russian immigrant students, being Russian meant being Jewish, even though they did not practice the Jewish religion—immigrant Russian youth identified with being of Jewish culture. Race, ethnicity, nationality, and religion, rather than shared affinities across those categories, became dominant markers of identity for the Redivided Generation.

Because student groups are inextricably linked to notions of individual identity, more research is needed to explore the influences of multicultural curricula on students' identity constructions over time. Toward the end of the previous century, it is more likely that native-born public school students were exposed to the histories and points of view of U.S. residents from different racial, class, and ethnic perspectives. Perhaps graduates of the Redivided Generation were schooled in dividing the world according to categories of race, class, ethnicity, gender, and so on; and schooled in seeing each other belonging to one of those groups. Perhaps the self-segregating tendencies of the Redivided Generation reflect institutional influences more than demographic effects.[6] Still, school policies and institutional norms exacerbated the nefarious effects of overcrowding and the disorienting effects of accelerated diversification at Miller High during the last decade of the twentieth century.

There is reason to believe that student loyalty to racial and national groups were intensified by the frequency and severity of punishment for transgressions. Vera Debin and Ivan Strasky remembered how they stood up for their Russian peers, even when they risked getting in trouble themselves:

> Vera Debin: "Somebody said something wrong, or pushed you out of the way...If somebody would say something in front of a Russian student, the other Russian students would kind of back him up. So, it would be scary."
>
> Ivan Strasky: "There have been a lot of times—'Oh, that black kid said something about the Russian kid'—and then, 'You guys, after school, parking lot.' Countless times, and I'd go, because in the back of my mind, 'Ivan, what are you going to do? You're not going to fight, because if you get in trouble, it's not worth it. But if something bad does happen, you have to stand up.' Even in my academic mind, when I knew what I had to do to succeed and go to college, I still had this association...that I have to go."

Judging by Ivan's testimony, one can easily extend the same kind of loyalty to the black students who engaged in the parking lot fights that Ivan described, as well as to the "redneck" students with KKK tendencies. While seeking comfort might have primarily propelled ex-city blacks and Russians to retreat into their own groups of black and Russian students, zero-tolerance policies might have contributed to sealing and consolidating group loyalties around racial and ethnic boundaries. Loyalty to racial or ethnic groups might further explain why young county blacks would more easily connect with city blacks, and why the mother of a white boy assaulted by a Russian boy might have assumed, as earlier mentioned, that "the rest of the Russian kids would come to her house and beat her boy up" out of loyalty to their Russian peer.

Rigid school policies of the 1990s contrasted dramatically with those of the Border-Crossing Generation. In the 1970s and early 1980s, when students knew each other from attending the same elementary schools, when the population had remained fairly stable, and when school policies had been relatively lax in comparison to zero-tolerance policies, students experienced greater interracial and cross-class friendships and greater intermingling across categories of race and class. During the 1970s, the Border-Crossing Generation also basked in the broader lingering effects of the civil rights movements. But by the 1980s, as Tyack and Cuban suggest, "Policy talk about schools stressed a struggle for national survival and international competition...In such periods, policy elites want to challenge the talented, stress the academic basics and press for greater coherence and discipline in education." The authors remark that by the close of the twentieth century, in the "late 80s and 90s...conservatives and liberals alike...called for national standards."[7]

The contested institutionalized nature of the preps, the one group of students who continued to construct the public image of the school across time periods, might have further contributed to lessening student encounters across groups. While stories of alumni who graduated in the 1950s suggested that students collaborated with preps and participated in school social events across class divides—even across race divides in the later years of desegregation in the 1960s—and while the stories from the Border-Crossing Generation suggested a tolerated prep contingency and a continued participation in prep-organized events, albeit not without criticism, stories from the Redivided Generation revealed outright rejection of preps through prom boycotts and moral indictments of their values, as earlier discussed. The ironic and unintended consequence of student boycotts of prep-orchestrated social events might have been that they also robbed students of opportunities to meet across social divides. Neither recollections nor yearbook data suggest that alternate groups of students might have taken charge of creating social gatherings parallel to those of the preps. In fact, it was Vera Debin, a critic of the preps, who underscored that "[t]hey are the people with the school spirit. I can see why they are representing the school." Needless to say, it would have been easier for native-born youth whose white middle- and upper-middle-class backgrounds aligned with school authorities to feel most empowered at Miller High and least challenged or suspected by teachers and administrators.

The Class Issue

The upper-class economic status of white, black, and Russian preps would have plausibly further galvanized the frustrations of those students identified as "farmers" or "rednecks," as they felt snubbed by preps, who by the end of the century included black students. The reader will recall lower-class white students' refusal to recognize Nat Right's middle-class status in the 1960s because he was black. The reader will also recall Cherry's disdain, in the 1990s, for the preps, Cherry who identified herself as "a back-home farm-country girl." These, among other testimonies earlier discussed, further suggest that the undercurrent of animosity expressed by some of those white students identified as "farmers" and "rednecks" toward the visibly rich and toward blacks, an undercurrent that might have been repressed in the 1970s and early 1980s by vigilant white peers, surfaced and exploded in the 1990s.

The preps of the 1990s came to represent a coalition of races and ethnicities bound by superior consumer power. For the "rednecks" of the community, who witnessed the disappearance of the farms they had worked on for generations as the general consumer lifestyles of the city took over, this coalition might have seemed overwhelming, and more than that, as recollections seem to suggest, was an affront to their "farm country," "down-home" way of life.

Finally, that graduates of the Redivided Generation were remembered as rule abiders or rule breakers suggests an environment more akin to that of inmates in a prison than of students in a school. Remembrances of an explosive cafeteria where racial fights erupted and where racial and ethnic threats and slurs were exchanged describe the atmosphere of a prison mess hall rather than that of a place for students to congregate in safety and to eat. There is little doubt that school authorities were taken aback, in the late 1980s and 1990s, with the sudden influx of youth, both immigrant and from the city, as they scrambled to build an annex. With the privilege of hindsight, however, this historical analysis shows that segregating forces had already been entrenched in the organizational structure of Miller High, which fully came to view with the avalanche of newcomers. Structurally inherited tracking divisions continued to isolate students into parallel worlds[8] and to expose students in lower tracks to violence more readily than students in the upper tracks. A cafeteria devoid of teachers' and administrators' involvement and role modeling allowed for racial and ethnic hatreds to erupt.

Youth and Authority

At the end of the 1990s, the highly competitive climate for access to colleges increased student rivalry in the upper tracks. As a result, student-to-student relationships hardened both within and across tracks. Achievers competed against each other, sometimes ferociously, as Sue

Cohen's testimony regarding the aspiring valedictorian revealed, and as suggested by the recollections of Bill Jackman, Ivan Strasky, and Vera Debin. In the 1950s and 1960s, white students easily coasted through upper tracks in large part because the upper tracks were economically rather than academically defined, and were by and large reserved for those who could afford college tuition. Those most likely to work hard in the upper academic tracks were black youth. In the 1970s and early 1980s, students across race and class invested minimal effort in school, particularly because the older teachers held low standards for their charges. In general, students of the 1970s and early 1980s were less invested in studies or in pleasing teachers. By contrast, the Miller High upper-track student in the Redivided Generation felt the pressures of academic competition as more middle-class youth competed for choice colleges, and as credentialism created expectations for universal college attendance. Of the previous generations' students, perhaps the experiences of 1950s graduate Norman Good, the African American alumnus who attended the academic track where he had to prove his competitive spirit, might have approximated those of the 1990s academic-track students. Still, as Norman Good recalled, he "competed in a good-spirited way for grades." Memories of 1990s graduates revealed how rather than "good-spirited," competition was strategic and stressful.

Moreover, achievers constructed their upper-track status as proof of their superior qualifications by pointing out that students attending standard classes were either of lesser intelligence, as Ivan's recollections emphatically suggested, or not as hardworking, as Bill's and Vera's testimonies revealed. In this sense, graduates' recollections echoed those of the Divided Generation who constructed "lower-track" students as less deserving.

When Miller High students focused on ensuring their particular student status rather than their peer status, as was the case for both the Divided and Redivided generations, then disdain by upper-track attendees for lower-track attendees intensified. Within a dividing and hierarchical system, the more teacher-centered the students, the more likely they were to construct each other hierarchically, according to institutional divisions. The reader will recall how in the 1970s and early 1980s, teachers' divisions of student status according to tracks were not embraced by the Border-Crossing Generation, whose respect for teachers had plummeted when compared to that of the Divided Generation in the 1950s and 1960s.

What is of importance here is that teacher-centered students—students invested in either pleasing their teachers (1950s and 1960s), or befriending them as they sought competitive advantages (achievers of the late

1980s and 1990s), or not crossing them to ensure protection in a zero-tolerance climate (rule-abiding standard-class attendees of the late 1980s and 1990s)—integrated notions of one another as superior or inferior as defined by classes taken and grades acquired. Thus Ivan and those students institutionally recognized as belonging to that "pole" of Miller High youth whose academic status set them apart from the majority of students considered someone like Cherry, who attended the general classes, of lesser merit. Cherry, in turn, placed herself hierarchically lower than GT attendees Ivan, Sue, Bill, or Vera. Recall how Cherry, Heather, and Roberta situated their place in school somewhere between "the group that did drugs" and the "higher group...that you knew were really smart." Tellingly, as if in an attempt to avoid the lowest status along academic tracking, they placed themselves above those who "did drugs," a qualification that does not define one academically but socially. Also telling, not doing drugs accorded them a status of good conduct, which in turn ensured acceptance and belonging to an institution hypervigilant against transgressions and readily apt to expel.

However, while the Redivided Generation resembled the Divided Generation in that it was school authority centered rather than peer centered, students of the Redivided Generation did not develop authentic personal relationships with school authorities as did graduates of the 1950s and 1960s. Rule-abiding students of the Redivided Generation objectified or abstracted school authorities. Achievers saw them as a "means to an end." Recall Ivan's words: "Kids at that level always have great relationships with teachers because they do whatever it takes to get the teacher to like you...To get an A you need a relationship with the teacher." Standard-class attendees saw school authorities as ivory-tower keepers of the law whose presence was made manifest only through expulsions and punitive actions. One alumnus's recollection summed it up: "I cannot remember seeing the principal walk the hallways...You stayed away from the office."

Rule breakers also objectified school authorities and continued to clash with teachers, whom they constructed as "the system." However, unlike the rebels of the 1950s and 1960s, who could be saved from endless detentions by a shop teacher taking personal interest in their lives, rule breakers of the Redivided Generation were less likely to be pardoned and were also less likely to develop meaningful personal relationships with teachers. Meaningful relationships with teachers were also less likely for successful students and those fully integrated into the life of the school. GT student Vera Debin recalled: "There was a very few teachers that I had any kind of relationship with other than in class," and the prep jock Cecilia Hood said: "I don't remember the teachers that much." Graduates of the Redivided

Generation talked about teachers in abstracted terms, and none of those interviewed spontaneously recalled or named a particular teacher as had graduates of the 1950s and 1960s—further underscoring the segregation between teachers and students, and between administrators and Miller High youth by the end of the century.

The cafeteria continued to be a herding place where students were left to their own devices to establish relationships in isolation from their teachers, who ate in their own quarters. Students, segregated from their teachers during mealtime—a time notoriously reserved, across cultures, for community and communal relationships—further segregated from one another as they sought familiarity and comfort. The teacher-student segregation vividly came to light in Ivan's description of his privileged position with teachers in a school system where relationships are constructed hierarchically. When Ivan said, "I was president of my class, I'd always be in the lunchroom with the teachers," he made it clear that he was not part of the student cafeteria crowd, but that he had access to the private world of faculty who ate away from the students who were considered lower in the hierarchy of relationships.

It is little wonder then that within the space of the cafeteria, racial, ethnic, and class animosities flared up in the overcrowded Miller High of the 1990s. Carnoy and Levin's remark that "schools continue to provide Americans with a social experience that is markedly more egalitarian and more open to free choice and possibilities of self-realization than anything that is available to them in the realm of work"—a remark that might have captured to some extent the experiential realities of some students of the previous two generations—no longer applied to the Redivided Generation. By the end of the twentieth century, Miller High was a place where students participated less and less in the life of their school, and where competitions within upper and lower tracks were as stressful as those found in the world outside: Achievers competed more fiercely to keep ahead; rule-abiding standard-class students held their breath hoping not to be sucked into what they perceived to be growing violence around them; and rule breakers broke rules with greater consequence than ever before as they were apt to be judged in courts as adults.

A Student Body that Negotiated Discrimination but Remained Politically Inactive

Native-born white graduates Bill Jackman, a Christian conservative youth when he attended Miller High, and Heather Korn and Cherry Gate, rule-abiding conservative standard-class students, reframed discussions around discrimination also to highlight "reverse discrimination"—discrimination against whites by blacks. Jackman spoke of people in school "that were racist either way." Heather Korn complained about privileges accorded black youth that were denied to white youth.

Heather's recollection spoke of unfair treatment against whites and of their disempowerment as they were just "sitting back and watching this happen." The reader will also recall Cherry Gate's fear of black students as she recalled death threats against her sister implying black racism against whites at Miller High. Some white Russian youth also perceived that they were discriminated against by "American" schoolteachers just because they were Russian. But Ivan Strasky further suggested that many were quick to call discrimination in situations that in fact were teachers' standards for academic performance for all students:

> Whenever the teacher challenged them, not because of being Russian, but because the teacher was being a teacher—"Why didn't you do your homework? Why are you late?"—they would automatically be defensive, and often perceive it as discrimination.

Recollections suggested that segments of native conservative white youth and Russian immigrant youth negotiated the idea of discrimination to underscore their victimization by blacks or by the "American" educational system, thus suggesting a student body fragmented into communities that were hypervigilant of discrimination against their own groups' racial, ethnic, national identities. Defensiveness and defense of group identity along racial, ethnic/national divides characterized this generation. Of interest is that the Redivided Generation did not recall self-segregation on the part of students along gender divisions as had the previous two generations, or discrimination by school authorities along gender divisions as had the Divided Generation of the 1950s and 1960s—perhaps suggesting a more broadly shared sense of gender equality by the end of the century.

This hypervigilance and hyperdefensiveness, particularly along racial identities, comes even more clearly into view when compared to previous generations' racial relationships, and even more so when comparing young women's experiences in the lower tracks. In the 1950s and 1960s, racial border crossings between young women were timid. The reader will recall how African American alumna Doris Right, who graduated in 1958, having transferred to Miller High as a junior from the all-black school she had been attending prior to desegregation in 1956, described a white girl's tentative attempt at making friends.

Twenty-three years later, the description of racial border crossings between African American and white adolescent women would prove more daring, but also more dangerous for African American young women. The reader will recall 1981 graduate Theresa Randle's experience with her friend's KKK father.

By the mid 1990s, however, stories emerged of racial fights between young women.

Young women's relationships across race in lower tracks progressed at first from timid to daring between the Divided and the Border-Crossing Generation, then shifted into racial violence by the end of the century. Until the Redivided Generation, young women's interracial relationships had been more conciliatory than young men's. This reversal in trend among young women in lower tracks toward greater interracial relationships may be in great part attributable to institutional messages about race relations transmitted across time periods.

During the early years of desegregation (1956–1969), the institutional message regarding race relations was that black students should overlook racism and be patient. During the Border-Crossing period (1970–1985), the institutional message was that the efforts for inclusion should be made by black students, even as many youth crossed racial, gender, and class borders without much reference to school or parental authorities. The reader will again recall the testimony by African American alumnus David Randle, a 1976 graduate: "[N]obody approached us about [joining sports]...so what happened...a lot of kids they didn't get involved in anything."

During the Redivided period (1986–2000), the message was that the responsibility for violent racial altercations should be equally shared by white and black students, as Heather Korn's testimony suggested.

From "bear the racist insults" in the 1950s, to "find your own way into mainstream life" in the 1970s, to "assume equal responsibility for racial violence" in the 1990s, the institutional messages across generations consistently betrayed a "washing of the hands" by school authorities of any responsibility for helping youth integrate authentically across racial and other social divides—authentic integration understood as "intellectual and social engagement across racial and ethnic groups,"[9] as addressed in the Introduction. It is not surprising, then, in light of continued lack of adult positive involvement in the creation of an authentically integrated school climate, that youth redivided. Consistently, the onus for integration was placed directly on the shoulders of students. The students, in turn, by the end of the century, were ready for battle, whether along racial or nationalistic divides. From complaints about teachers' unfair treatment to actual racist peer behavior, students readily voiced accusations of discrimination and fought racial fights.

Evidence suggests that just as the two preceding generations, the Redivided Generation was not politically involved. The balkanization of students within an overcrowded school environment, groups of students with adversarial dispositions toward the U.S. educational system and the U.S. government more broadly, and a segment of the student population whose energies were consumed by commitments to scholastic success above all else might have contributed to a more narrow, immediate, and personal focus on part of students. Then, too, students who decried the consumer-oriented factions of Miller High and who might have found a basis for political action around issues of consumerism and class inequity were students whose group affiliations were oppositional: Russian, "redneck," conservative Christian, and punks expressed their affiliation aesthetically rather than politically.

* * *

The insular and xenophobic tendencies of "rednecks" and their long-standing racism; the mistrust and survival skills imported by newly transplanted black city youth, and by many Russian youth eager to change their fortune in "America," all the while begrudging it its Americanism; the zero-tolerance policies that divided those who toed the line and those susceptible to expulsions; the explosive overcrowding; and the "prep factor" combined to create strands of clashing identities claiming their space and legitimacy within the halls of Miller High, and within an educational system that by the end of the twentieth century was scrambling, yet again, to reclaim its relevance.[10] After forty-four years under *Brown v. Board of Education*, Miller High students segregated according to adult-devised census categories of identity, and lived parallel lives further divided from those of school authorities.

Conclusion

Authentic integration at Miller High over the second half of the twentieth century had been the experience of too few students across generations. Interracial friendships that lasted beyond school years were mostly forged during the 1970s and early 1980s, an anomaly in this fifty-year history. A propitious convergence of social forces sustained the Border-Crossing Generation: a broadly shared middle-class status across racial divides; a population of black Miller High students who were both minorities and mostly children of families long established in the region; a student body accustomed to desegregated institutional settings; and a national consciousness of civil rights that stirred the imagination of an otherwise conservative student population in a conservative town.

Structural Forces

One may further claim with some confidence that without the national backdrop of the civil rights movement, the Border-Crossing Generation would not have forged relationships across divides to the extent it had. Local conservative politics mitigated the influences of broader national forces. Blacks with roots in the community that extended back to the 1800s, whose small businesses received white patronage, and who were "familiar with getting along with white people," and whites in the broader community for whom the status quo was comfortable, together created a local context within which Miller High youth would not engage politically to demand curricular, extracurricular, or other changes that peers in other parts of the country were claiming. The Border-Crossing Generation democratized peer relationships despite institutional norms but while not challenging them. The result, however, was that a segment of students—disenfranchised lower-class white youth and marginalized black students—continued to segregate during the most democratic period in the history of Miller High. The social context that proved favorable for nurturing cross-racial friendships, in particular the broadly shared middle-class status,[1] and the socially emancipating effects of the civil rights movement on black and white students, might also have been the factors that kept poorer whites and culturally critical young blacks of Miller Town segregated—the poorer white students because they were economically marginalized in a predominantly middle-class world; and the critical black youth, because they were culturally marginalized in the overall conservative Miller Town of their times. In the midst of Miller High's most authentically integrated period, segregation of poorer whites and disengaged blacks continued, underscoring the perdurability of segregation among students in Miller High over time.

Across time periods, too, cross-class friendships within and across racial divides were also rare. For each generation, they occurred mostly in upper academic tracks and a few extracurricular activities, most notably sports fields. These spaces were the least populated and most insular—spaces for the chosen few. Poorer students' possibilities for participation in upper academic tracks or extracurricular activities, which proved central to developing relationships with peers across diverse backgrounds, were restricted. Those students of poorer backgrounds who could join the upper academic tracks or play in organized sports increased their chances of forging friendships across class as well as race divides. Within elite spaces, encounters across differences were more likely to be friendly. Within the more populous general and vocational tracks, on the other hand, they were more likely to be unfriendly.

Indeed, the general and vocational tracks remained overall divisive spaces and incubators of racial violence. More important, animosities were not only reproduced, but also intensified within these tracks. In the confined spaces of Miller High, students could not avoid each other by dispersing into alternate realities as they might have in the larger community. The cafeteria also remained a place where students continued to segregate as they sought comfort in the familiar, demonstrated loyalty to their own, and consistently were left by school authorities to fend for themselves. During the 1990s, when demographic shifts brought into school perimeters Russian immigrants and black city youth from segregated schools and mistrustful of the school system and of each other, spaces such as the cafeteria only exacerbated their segregating tendencies[2] as they sought refuge in familiar ethnic, racial, religious, and economic relationships.[3]

Across time periods, too, changes in student relationships and demographic configurations were accompanied by changes in disciplinary measures and types of teacher-student relationships. Although a clean, linear progression and one-to-one correspondence between disciplinary measures, demographics, and student relationships across generations cannot be established, given the evidence, it is reasonably safe to advance that overall, as the student population grew and diversified, school authorities became more distant and disciplinary measures became more severe. Evidence also suggests that zero-tolerance policies at the end of the century might have exacerbated student segregation by consolidating and hardening student loyalties along racial and national lines.

The case of Miller High undermines the assumption that desegregated comprehensive schools have actually lessened racial prejudice[4] since *Brown v. Board*. First, this history challenges Grant's affirmation of an achieved "genuine racial equality." By extending educational historian Gerald Grant's story of Hamilton High beyond the mid 1980s into the end of the twentieth century, this history reveals that Grant's affirmation holds only under certain demographic conditions. Similarly, this analysis complicates the findings of a recent groundbreaking mixed-methods educational study, *How Desegregation Changed Us: The Effects of Racially Mixed Schools on Students and Society,* which examined the experiences in several desegregated schools of graduates of the class of 1980.[5] The study findings suggest that "school desegregation fundamentally changed the people who lived through it...desegregation made the vast majority of the students who attended these schools less racially prejudiced and more comfortable around people of different backgrounds." The findings in *How Desegregation Changed Us* confirm some of the findings in this history: Those who attended high school in the 1970s and early 1980s (Border-Crossing Generation) more readily forged friendships across racial divides that lasted throughout their lives. However, this history has further revealed that for most students such experiences did not continue into the end of the twentieth century. More important, this analysis has shown that across time periods, students in the general and vocational tracks did not experience desegregation in positive ways. The authors of *How Desegregation Changed Us* interviewed people who had experienced desegregation twenty-five years ago, when the nation rode the crest of the civil rights movement, and interviewed "educators, advocates, and local policy makers," a population of people less likely to have attended the vocational and lower academic tracks.

Experiences

This analysis also shows that by the end of the twentieth century, Miller High no longer provided its students with experiences "markedly more egalitarian and more open to free choice and possibilities of self-realization than anything that [was] available to them in the realm of work," as Carnoy and Levin have suggested. While middle-class white women in the 1950s and 1960s, and middle-class black women of the 1970s would be shocked at the discrimination they encountered in the workplace upon graduation, evidence of their more egalitarian experiences within the confines of Miller High, students of the 1990s experienced Miller High *as* the workplace. By the end of the twentieth century, fewer students participated in the social life of their school than ever before in its history. Those attending the upper academic tracks competed more fiercely, and developed instrumental, professional-like relationships with teachers as they negotiated their way to the top. College attendance having had become de rigueur—these upper-track youth engaged in competitions almost as vicious as those found in the marketplace. On the other end, rule-abiding standard-class attendees welcomed security measures they felt protected them against what they perceived to be escalating violence around them; and rule breakers broke rules with greater consequence than ever before, expelled if need be for their offenses and sentenced as adults in courts.[6] While to a greater or lesser extent Miller High had always served as a holding place for many of its youth, in particular male youth, over the fifty years of this history, this role intensified with the unwitting consequence of increasing (rather than lessening) social unrest and violence among many students, even as it provided a few safe spaces for a select few to forge cross-racial and cross-ethnic relationships.

On an intimate and individual level, there are the darker and brighter stories that each generation told. The darker story of the Divided Generation of the 1950s and 1960s told of the burden of early desegregation borne by local black youth assaulted by white youth, and by young black women who were further subjected to sexual harassment by white males; of the humiliations of the very poor and of pregnant women across race; of the alienation of male students inept at studies; of the disproportionately privileged lives of white middle-class young women when compared to those of black middle-class young women, and of their favored status when compared to all other students. On the other hand, the brighter story of the Divided Generation told of young black and white women's impulses to cross over divides against entrenched racial mores of the early days of desegregation; and of black and white neighborhood friends having lunch together within the spaces of the cafeteria while blacks still purchased goods from the back doors of the community's white-owned stores; of young black men's academic triumphs and relationships forged across racial and class divides; of poor whites finding their way into middle class; of black women gaining momentum, by the mid 1960s, on their white counterparts and using their high school education to pursue professional degrees in higher education.

The darker story of the Border-Crossing Generation of the 1970s and early 1980s told of marginalized black and work-release youth who were ignored by school authorities; of the continued humiliation of the very poor; of coaches' abusive ways against young men; the egregious affairs of some male teachers with their female students; of the privileged status accorded by school authorities to jocks and the smart students at the expense of most other students across racial divides; and the humiliating labeling of youth by teachers as "losers." The brighter story of the Border-Crossing Generation told of friendships sealed between white and black young women and men on sports fields, in music bands, and classrooms; of interracial dances and couples; of white youth's vigilance against racism and commitments to egalitarian relationships across class and race boundaries; and of the social mobility afforded black youth through life-lasting connections established with wealthier white friends.

The darker side of the Redivided Generation of the late 1980s and 1990s told of student rivalries and prejudices along racial, class, ethnic, national, and religious divides; of fights erupting in the cafeteria and parking lots between Russian, black, and white racist youth, female and male; of stressful and fierce competition in the upper academic tracks for admission to colleges; of ready accusations of discrimination against teachers by students, and against students by students; and of broader community segregations that pitted old-time black and white families against newcomers to Miller Town. The brighter story of the Redivided Generation told of substantially increased social mobility for many African American youth whose economic status gained parity with the wealthiest whites in the community; of continued interracial friendships forged on sports fields; and of interracial and cross-national friendships forged among the wealthier preps.

* * *

Linking the Particular to the More General: Point and Counterpoint

That this history has unearthed divisive tendencies of the high school is not new knowledge. The comprehensive high school's systematic sorting of students has been well documented in educational histories and ethnographies, often linked to broader capitalist and market economic forces and examined in terms of institutionalized racism.[7] That this history has further revealed the role played by students in importing cultural influences from outside of school is also not knew knowledge, but is a phenomenon explored by anthropologists and educational ethnographers, although barely by historians of education.[8]

While this analysis confirms findings regarding the nefarious effects of tracking and students' economic and cultural realities imported into school spaces, it further expands our historical knowledge of student life and intergroup relations in high school beyond the historical implications already discussed in this conclusion, and beyond this work's extension of Gerald Grant's story into the end of the twentieth century.

First, this history reveals the impact of institutional norms on student relationships across dimensions of race, class, gender, and ethnicity, rather than on student achievement across those dimensions. As one researcher noted, "[W]hile student achievement data are critical, schools do more than teach academic subject matter: They have a profound potential for shaping individuals and their social networks."[9] In turn, social networks might play an equal if not more important role than achievement in post-school life opportunities. Jencks et al. found that "academic achievement does not have the kind of overwhelming impact on later occupational success that might justify making it the exclusive focus of most research on the outcome of desegregated schooling."[10]

Secondly, because this history brings to light the interplay across a substantial half-century, between school, demographics, social context, and citizens' lived experiences, it reveals across time institutional areas of intensified segregation and of facilitated integration": places of greatest vulnerabilities to divisive social forces and of greatest capacity for fostering authentic integration independent of broader social forces. Across the past half-century, as earlier discussed, Miller High's areas of intensified segregation proved to be the lower academic tracks and the large gathering space of the cafeteria, where conflicts heated up. Miller High's areas of integration proved to be the upper academic tracks and sports fields, from which students emerged into adulthood with habits of association that predisposed them to bridge divides—spots removed from heated conflicts. Miller High's tracking system, whether understood critically as a capitalist social-control mechanism and the tool of institutionalized racism, or benignly as the high school's democratic effort to "serve the needs and interests of all in a diversified curriculum,"[11] created school spaces that more readily divided or united students.

Historians have suggested that in the early twentieth century, high schools strove "to assimilate and amalgamate [the myriad diverse youth]...to Anglo-Saxon conception[s] of righteousness, law and order, and popular government"; that they "socialized youth to modern bureaucracy and connected their most intimate identifications to the power of the state"; that, in sum, they "allowed students to imagine that they were part of a national community."[12] The case of Miller High shows that by the end of the twentieth century, the comprehensive high school lost this purchase to amalgamate the students in its charge (if it ever truly had it—a subject for another study). This comes to view most vividly in the rejection of the "American way" by segments of the Russian and African American students, and in the racist attitudes of segments of native white students as well as segments of the Russian immigrant youth.

Under a half-century of changing demographics and a progressively more complex cultural environment, Miller High continued essentially to operate as its early twentieth-century counterpart. The case of Miller High suggests that while the comprehensive high school of the end of the twentieth century continued to track and sort citizens—business as usual—it became, within the new world order following desegregation, the technology that tightened divisions along religious, ethnic, and racial identities in ways that intensified animosities across divides, rather than fostered respect across those divides. Moreover, while the comprehensive high school of the early twentieth century had also been "the indispensable technology that helped reorder the relationships between self and society...[and] loosen the grip of local, religious, ethnic, and racial communities,"[13] the case of Miller High suggests that by the end of the past century, it became the very technology that splintered U.S. citizens and residents into diverse local loyalties.

By the end of the twentieth century, a very diverse student body had been sensitized to issues of discrimination, the case in point being Miller High's recently immigrated Russian youth who openly accused "American" teachers of discriminating against them, something that immigrant youth of the nineteenth century would have been less likely to do. Another case in point is Miller High's recently migrated city youth, who had been much more attuned to racism than the local black population. While at the turn of the twentieth century the comprehensive high school might have been a response to the perceived threat posed by diverse youth ready to compete for jobs with adults, and while it became the technology that kept youth off the streets and served to assimilate them into the order of the day, as Joseph Kett and other historians have suggested, by that time it had lost such authority. The early twentieth century "threat" of diversity had been dislocated from the street into the school by the end of the twentieth century. Diversity's "threat" to marketplace social order had developed over time into a threat to "school business as usual." Miller High's Anglo middle-class normative practices and its institutionalized regimentation and herding of student bodies turned "diversity" into a problem rather than a solution.

At the meeting place between a racially, ethnically, and religiously militant youth and the rigid institutional normative habits of tracking and herding, for school authorities, social order seemed threatened within school walls. Unimaginatively, school authorities responded by becoming more punitive. However, the reader will remember how by the 1990s rule-abiding students in lower tracks preferred strong autocratic disciplinary measures that made them "feel safer," protected against law-breaking students, who often happened to be Russian, black, and "rednecks." The reader will further recall how by the 1990s divisions among students created parallel worlds, a citizenry divided not only by race, ethnicity, and class, but also as rule breakers and rule abiders. One can reasonably claim that the dark side of this history presaged the rise of a citizenry accustomed to a policing government. The case of Miller High shows a punitive system sustained by the fears of rule-abiding Miller High students, further fueled by aggressive and mistrustful disenfranchised students and a rigidly divisive tracking system.

Miller High had increasingly opted over time to police students—to keep order under business as usual. In considering this darker history, one might further venture to presage the future weakening of democracy, witnessed in the increased policing of citizens at the dawn of the twenty-first century (e.g., Patriot Act, wiretapping), a policing that has been examined only sporadically and with delayed reaction. One might also venture to presage, as we globalize the comprehensive high school, the making of a multiracial, multiethnic, economically advantaged minority (the multicultural preps), and a multiracial, multiethnic, economically disadvantaged majority, as we "stuff" more students into lower tracks. These sinister, science fiction—like notions, while too simplistically and too harshly forwarded in these concluding remarks, are not unimaginable if the comprehensive high school is to continue business as usual as it has since its inception.

There is, however, a counterpoint to the dark story. When looking at the brighter side of this history, captured in the Border-Crossing Generation, one can imagine a civil rights renaissance and a student body that once again reaches across divides. One can imagine a youth buttressed by a secondary public school system architecturally designed to create smaller spaces for students to interact across differences. One can imagine a vocational or apprentice system incorporated into the workplace and attended by students following graduation from high school and not during high school. One can thus imagine dismantling the "comprehensive" in the comprehensive public high school and opting instead for a liberal arts curriculum for a twenty-first century global reality. A shared liberal arts curriculum would do away with the low status lower academic tracks within which, as this history has shown, racial and ethnic divides are intensified. Moreover, a shared liberal arts curriculum might also help lessen class divides since lower tracks are mostly holding places for students of lower socio-economic background. Finally, one can imagine a constant presence by a diverse body of school authorities among youth, one that exemplifies relationships across divides in everyday interactions, rather than a policing, distant administration. One can then imagine a fresh generation of youth that gives birth anew to a U.S. democracy that best exemplifies a "world lived in common with others."

Addendum Methodology: The Transparent Historian

The purpose of this section is to describe and discuss the challenges for data collection and interpretation of what is known among historians engaged in collecting oral histories as "oral history doing."[1] I begin by addressing the use of the guiding theoretical framework intersectionality in collecting the oral histories for this analysis. Historians de rigueur abhor theory, and for valid reasons, but I argue, in what follows, for the relevance of this particular theoretical lens for this particular oral historical research. I then address the more technical aspects of oral history doing: identification and location of graduates, interview protocol, transcribing, and coding. I end by describing my archival research and the particular analysis of yearbooks for this history.

Theory and Oral Histories

I would like to begin by emphatically stating that oral historians not only reproblematized memory for historians by focusing attention on understanding the subjectivity of memory as a manifestation of historical consciousness,[2] but that they also brought to our attention how memories are gendered, racialized, and class based. They brought to our attention the importance of examining why different individuals and groups experience the same event in very different ways.[3] Today, oral sources—particularly since the 1990s, when historians began investigating the construction of identities[4]—are compared less pejoratively to documentary sources by academic disciplinarians. Additionally, oral historians continue to break down "boundaries between the educational institution and the world, between the [history] profession and ordinary people."[5] However, they also continue to be faced with the challenge of articulating "the connection between individual and social historical consciousness."[6]

This theoretical challenge—transforming individual memory into social memory—becomes particularly relevant to those of us interested in the production and analysis of oral testimonies of school life across generations in the United States, since desegregation, as these reflect on peer relationships across socially constructed categories of difference (race, class, gender, religion, nationality, etc.). When, how, and under what circumstances do individual memories of school relationships credibly tell a social rather than psychological story? How does one determine the number of testimonies necessary to legitimate a social rather than personal history?

To this theoretical challenge is added the *problématique* of the relationship between interviewer and interviewee, professional historian and narrator. When oral historians stir up memories of school experiences, they stir up worlds of identities in the making. They also stir up present-day integrated meanings of school and the educational system, meanings that simultaneously interlace, buttress, and challenge memories. Who speaks: the adult reconstructing the past to justify the present, to justify the past, and to whom? To herself or to the historian? How to deal with the power-laden nature of co-creating with the narrator a primary source—that is, the very data one uses to build one's analysis?

Fundamental to the theoretical and methodological challenges posed by researcher-participant co-constructed interviews is the issue of language itself. As a former foreign language teacher and avid student of sociolinguistics, I construct voices as social rather than primarily intrapsychic phenomena. In this view of language, meanings are constructed by individuals as these individuals relate to one another.

"The word in language is half someone else's... It is populated—overpopulated—with the intentions of others... to understand voice, researchers must accept that what they hear is a function of who they are as individuals within the social community."[7] Thus the oral historian contends in her métier with Bakhtin's heteroglossia; that is, she navigates the many voices situated within dialects, generations, class, race, and national consciousness; she contends with the diversity of voices as they speak against or to each other, and as they speak differently or similarly to her own. The integrity of her work rests not only on examining critically the linguistic register of collected oral testimonies as well as archival documents, but also on critically examining the sociocultural intentions revealed by her own linguistic markers. The essential nature of oral historical inquiry naturally and relentlessly brings the oral historian to theory.

Broadly, theoretical frameworks provide needed discursive templates to talk about the challenges presented by oral historical work that seeks to examine how diverse students adapt to multiple cultural contexts and affiliations within single institutions over time. One of the top challenges is the obvious intrusion of the historian in the production of oral histories. The consequences of that intrusion often create metadata that cannot be easily tucked into the narrative, but that should not be relegated to footnotes. These are data that interrupt the narrative. I call metadata those bits of information that do not speak directly to the past that I seek to recover, but to narrators' present reactions to and reflections about the oral historical research process. These reactions often shed important light on narrators' developed historical consciousness about the story being researched. Theory helps organize human phenomena in their relation to one another. In the case of my work, intersectionality theory becomes a useful lens for the complex stories that I have to tell.

Black feminist theorist Patricia Collins's intersectionality theory offers a framework for positioning both historian and narrator within the "doing" of oral history. It also offers a framework for capturing alumni's shifting constructions of cross-group relationships within and across time periods.[8] The intersectional analytic approach examines the ways in which social markers of difference (race, gender, ethnicity, sexuality, generation, class, religion, and nationality) intersect to shape situated experiences, and places the historian within her own social position. Intersectionality challenges homogeneous constructions of social groups and brings to light the complexity and fluidity of group identities. The theory grew out of black feminists' articulation of the limitations of "gender as a single analytical category";[9] and by extension, the limitations of any single social category to capture the complexity of cross-group relations.

My application of the theory addresses both the intercategorical and intracategorical dimensions of intersectionality.[10] The intercategorical dimension focuses attention on the relationships between social groups and identifies the power differentials at play in those relationships. The intracategorical dimension historicizes social categories and brings to light their shifting definitions by identifying people who "cross the boundaries of constructed categories."[11]

Intersectionality applied to the oral historical enterprise per force situates the oral historian not only socially, vis-à-vis individual narrators, but also historically, vis-à-vis changing understandings over time of the researcher's status and role. I begin then by situating myself to you, the reader, that you may consider the many implications of my arguments from your own situated perspective.

It is from the situated perspective of a naturalized U.S. citizen; of a white woman of hybrid cultural and diverse linguistic backgrounds whose race and sexual orientation accord her important social and political privileges in a society that favors whiteness and heterosexuality, while her gender and role as mother restrict her economic and professional opportunities;

and it is further from the perspective of a multidisciplinary scholar who was schooled in French Canada and Mexico prior to graduate work in U.S. universities that I engaged in conversations with diverse alumni who attended Miller High between 1950 and 2000.

In the history of Miller High, subjectivities created and recreated one another. Graduates' testimonies did not merge and rise into collective memories for each generation, but rather into shared interpretive frameworks regarding the nature of peer relationships across markers of difference. Thus from a mosaic of differently positioned individual experiences emerged the relational historical consciousness of Miller High generations. While demographic composition, institutional norms, and broader political and economic environments shaped the perimeters of possible student encounters across social categories for any given period; and while the voices of alumni reflected the general social grammar of their times, the history of cross-group high school relationships is grounded in graduates' testimonies—in individual, subjective experiences. Remembered testimonies were given primacy over documentary data; that is, archival data was used to contextualize testimonies, not the other way around.[12]

Furthermore, while I conducted archival research before, during, and following interviews to situate the school and its community, and to see if "facts" stated in testimonies had any grounding in broader social realities or official data, the archival research was not conducted to challenge or discount the memories shared, but to situate them more clearly when possible. The "truth status of individual stories became less important than the value they [were] trying to support in the telling"[13]—because what was sought was the lasting impact of school on the development of school relationships, and the impact of school relationships on perceptions of "others" as experienced by male and female graduates of diverse racial, ethnic, and class backgrounds.

For relational histories of education—such as this one, interpreted from Miller High alumni memories—no ultimate number of testimonies can be set to legitimate a social historical consciousness (over personal history). No number can be set because no exhaustive story of all possible configurations of student relationships can ever be told. It is because the "exhaustive" will forever be out of reach in relational histories of education that histories of whatever possible configurations must be told. They must be told because they begin our foray into the evolution and/or dissolution of intercultural, multicultural, multiracial, multiethnic, and religiously, economically, and nationally diverse educational spaces.[14]

Intersectionality theory offers a way to connect graduates' subjective interpretations of school relationships to their situated positions within both school structures and broader social contexts, and in reference to others' subjective interpretations. It is the very juxtaposition of differently situated perspectives within and across time periods that builds a relational historical consciousness. Thus the intersectional lens helps the oral historian identify and articulate how socially constructed markers of difference intersect within individual historical protagonists and how they shape nuanced group experiences that do not monolithically assign equal experiences to all blacks, all whites, all women, all men, and so on, within institutional bounds. It is precisely because this lens helps the oral historian identify human experience within intersecting categories of socially constructed identities that it also helps the oral historian to identify more clearly the fluidity across time of the very social construction of identity categories. Thus, rather than contribute to essentializing experience within social categories or to perpetuating categories of identity that are arbitrary, the intersectional theoretical lens helps more starkly bring to light how historically specific these constructs are. Intersectionality theory helped this oral historian frame and organize what continues to be a challenging history to grasp.

I repeat what I have already addressed in the Introduction: Informed by the theory of intersectionality and committed to an oral historical perspective that foregrounds the subjective experience of history—not to rewrite facts, but to gain a deeper understanding of how and why the same facts are reconstructed differently across historical protagonists—I forged ahead to find and tell the story of Miller High students' cross-group relationships from those perspectives recovered.

Identification and Location of Alumni

I identified graduates by searching senior pictures in Miller High yearbooks. While overall and broadly I could identify graduates in yearbooks by gender and race, and even to some extent by ethnicity through last names, I could not as easily identify them by economic status; serendipitously, I fell upon graduates of different economic backgrounds (more about class identification follows). When graduates identified through yearbooks could not be located in the alumni directory, I forged ahead until matches were made. It was particularly difficult to locate African American graduates of the 1950s through the 1980s, and in particular men, who rarely included their names in the directory, a telling story of its own.

Doors opened up for me into the African American community when I came across a history, in the town library, written by Louis Diggs, an

African American oral historian who had collected the testimonies of members of some of the oldest African American communities in Maryland.[15] I contacted the author, who suggested I speak with Annie Milligan, an octogenarian who had lived her entire life in the African American community that Miller High served. Mrs. Milligan took me into her world and connected me with African American Miller High graduates. It was her trust in me and her standing in her community that secured my interviews with African American alumni who attended between 1956 and the early 1980s.

Through Mrs. Milligan's intercession I gained invaluable insight into Miller High's black community of the 1950s through the 1980s, but I could not do the same for the younger generation of the 1990s. As discussed in the body of this work—while those alumni interviewed freely agreed to give their stories to me, the implications of my whiteness within a U.S. reality, in conjunction with my PhD candidacy in education at the time of the research, and the power differential inherent in the production of research interviews, created in the eyes of two potential African American narrators, 1990s graduates, a person to whom they would not want to grant an interview. The agency of these two alumni was recorded within the narrative. They refused to give their insights and interpretations to someone whom they understood to represent an educational establishment that has failed black students. Their political statement reconstructed the power differential on their behalf. Their refusal to tell of their experiences is a historic act that was recorded as clearly as remembered testimonies. It could have been omitted or—perhaps worse in this case—footnoted. Their refusal is part of the metadata that deepen and complete the history of Miller High student relationships, but inconveniently interrupt the narrative.

By adopting the intersectional theoretical framework and abiding by its logic, one risks exposing one's vulnerability as a researcher. Mine was exposed, and in the process, a historical defiance was recorded, one that must be understood positively—the agency of real people in the face of institutionalized and institutionally sanctioned educational research that lacks immediate transformative power.

My situated persona also elicited a response from one white alumnus, which brought to view a parallel story of "old-timers" and "newcomers" in the Miller High community. Tim Whittle, a white alumnus of lower socioeconomic status when attending Miller High between 1974 and 1981, expressed an opinion about how I might be perceived by the town's "old-timers" at the time of the interview:

> In this huge town, of the people who are here now—as my grandmother used to call them, the 'move-ins'—she wouldn't even talk to you! You

"move in" with an accent! [Tim imitates his grandmother's indignation, and laughs, razzing.]

In this case, it was not my whiteness, but my foreign accent that created an opportunity for the alumnus to make a statement about a social hierarchy of belonging. Tim's comment alerted me to the possibility of a parallel story—a story that emerged from Tim's reaction to my linguistic register. Indeed, it became the story of cross-generational strife in the community between long-established black and white Miller Town families and newly arrived immigrants and black youth from Baltimore throughout the 1990s. But in the live human encounter of the oral historical interview, a metadata set also emerged to complicate the story: the fact that the researcher's accent made her suspect. Intersectionality compels one to reveal the very encounters with narrators that tell of the import of the situated position of each in the making of historical evidence and interpretation. Such transparency serves my work well.

My goal was to identify an average of ten graduates per decade, and to include as much as possible equal proportions of male and female graduates of different racial and economic backgrounds who attended during the first and second half of each decade. I decided on ten graduates per decade given my resources at the time of the research. Since my intent was primarily to locate different perspectives and to capture meanings ascribed to relationships in school rather than investigate the veracity of a past occurrence, difference, not volume, mattered. Secondly, grounding my study deeply within a single institution made collecting a smaller number of oral histories much less problematic than if I had spread my study across numerous sites. Finally, I reiterate that I was not interested in producing a "collective" memory of Miller High students over the second half of the twentieth century. Undeniably, such an endeavor would have required "volume" and would have carried headaches of its own, since by the end of the century, fragmented perspectives (as this work has shown) would have challenged the notion of a "collective" memory. Seeking to construct a "collective memory" works best when the group whose memory is sought is homogeneous.

Although I had originally scheduled to interview forty-five graduates, I ultimately interviewed thirty-seven. Eight canceled and I ran out of resources to pursue further interviews. Beyond the Russian population of the late 1980s and 1990s, immigrant students also included Asian Indians and Latinos. They, however, represented a significantly smaller community than that of Russian immigrants at the time, and again, while it would have proven invaluable for this study to include their perspectives (the absence of which I sorely lament!), I did not have the resources to invest in locating and interviewing these populations of immigrants.

I began the data collection process by sending letters to identified graduates. I included an explanation of my project along with a stamped return envelope for their reply. Five out of the fifty people I had written to in my first batch of letters answered my mail; two declined to be interviewed, and those who answered were all graduates of the 1950s. Of the second batch, none responded. The three graduates of the 1950s who did accept proved to be invaluable contacts, as did Mrs. Milligan, through whom I contacted several African American alumni who then led me to other alumni, black and white. Thus I located graduates through purposeful sampling as well as snowballing.

Graduates of the 1950s were much easier to locate and much more eager to tell their stories, so much so that as a single researcher, I could not afford to interview them all. While I could locate later graduates, they were not as eager to be interviewed, being busy with work and raising children. This explains the disproportionate number of graduates, and of female graduates in particular, interviewed from the period before desegregation in the 1950s when compared to the other decades. However, their perspectives served to establish a baseline, a construction of student relationships among all white youth immediately preceding the desegregation of Miller High in 1956. Their perspectives allowed me to capture the already institutionalized segregating forces at Miller High.

I proceeded with interviews as soon as I could schedule them, even as I worked to locate others. In the early stages of data collection, interviewing and locating graduates to be interviewed proceeded in tandem. While I had planned to begin interviewing graduates of the 1950s and of the latter part of 1970s first, since they were located at the beginning and middle of the time period—to feel out the decades—my plan soon dissipated. I had to adapt to alumni's schedules and availability. Nevertheless, I was able to spread out my beginning interviews across the decades and to pick up somewhat on changes in mood across time periods.

Interviewing Protocol

After introducing myself and explaining the project, I took time to make small talk while I prepared my recording equipment and to set narrators at ease about the taping process. I emphasized that we could stop the recording at any time (for a break, if something needed to be said "off the record," etc.), and I explained that they could choose to donate their recording to the town library now or posthumously, but that I would keep their identities confidential for the purposes of my work. Because I had previously conducted pilot interviews for which graduates requested anonymity, I learned that it would be best that all participants sign a letter of consent that underscored the confidentiality of their testimonies for the purposes of my study. Too many graduates are still active members of the Miller Town community and did not want to compromise their current relationships with comments made about the past that would become public. Some have agreed to donate their recordings to the historic room of Miller Town's library posthumously. To protect agreements of confidentiality, all proper names, including the name of the town and school, have been changed to pseudonyms. After long deliberations, I was satisfied that situating Miller Town within Baltimore County and Maryland would not compromise graduates' identities, since there are several high schools comparable to Miller High in the county, and since the influx of Russian and other immigrants affected several high schools, not only Miller High.

Thus, beyond an explanation of the purpose and procedures of the study, the consent form also included the following:

> All information collected in this study is confidential to the extent permitted by law. I understand that the data I provide will be grouped with data others provide for reporting and presentation and that my name will not be used. I will be given the option to donate my tape recording to the [name of town] library archives. If I wish to do so, I will sign a secondary release form for this purpose.

I began all interviews with open-ended questions that asked graduates to describe their relationships with peers, teachers, and authority figures. This proved a very productive way into the conversation for all my interviews because it provided the space for graduates to jump into their memories as these came to them, without my imposing a prior order. As graduates' comfort levels increased, I followed with more probing questions about details of their experiences and the meanings they ascribed to their high school education and diploma. Only twice did such a broad, open-ended question spur graduates to meander on tangents into their personal lives, tangents that had nothing to do with their high school experiences. Nevertheless, allowing them to meander helped establish a rapport that yielded fecund interviews. While some researchers send a list of interview questions in advance to help the informants prepare for the interview, I chose not to send questions in advance. I wanted the conversation to emerge spontaneously, and I wanted to avoid formally written questions that might restrict either alumni or me.

Thirty-two interviews were held in graduates' homes, mostly during weekends or in the evening after work; one interview was conducted in the lunchroom at an alumna's workplace; three were conducted in a quiet area of the archives in the town library; and one was held at a bookstore. I always gave graduates the choice of our meeting place; I also took precautions, especially when interviews were conducted in homes and at night, by arranging to be picked up by a family member. Overwhelmingly, graduates and I quickly established good rapport.

Tape-recorded sessions ranged from one to two hours. On several occasions I followed up formal interviews with informal chats by phone or e-mail. To the thirty-seven oral histories conducted with alumni, I added one group interview with three Miller High teachers. One teacher, a female coach, had been working at Miller High for thirty years. Between them, the Spanish and English teachers had fifty years of tenure. The opening question I posed was to describe their experiences of Miller High students over time. This interview lasted fifty minutes and was conducted between classes in a quiet room of the Miller High library. Finally, I also interviewed, but did not record, the principal. The meeting lasted half an hour. See Tables A.1 and A.2 at the end of this section for a list of interviewees by time periods.

About Class Identification

The social constructs "middle class," "poor," and "working class" are tricky categories within a U.S. reality, and even more tricky when one crosses racial borders. What constitutes a middle-class range for a white citizen might not constitute the same range for a black citizen across historical periods and geographic contexts. It is important to underscore that I did not a priori choose a sociological definition of class as I engaged in my interviews. Class identifications emerged organically as narrators situated themselves and/or others they remembered in contrast to their positions on a continuum of "poor" to "middle class." I reported narrators' economic status as they described it. Nine narrators did not identify what their economic status was when they attended high school. In those cases I broadly deduced their class backgrounds by identifying their childhood homes in the town, the jobs they entered following

Table A.1 Pseudonyms arranged by generations, graduation dates, and interview dates

Divided Generation (1950–1969)	Divided Generation (1950–1969)
Before Desegregation (1950–1956)	**After Desegregation (1956–1969)**
Judy Law (1950–1954) / 02–25–2003	Doris Right (1956–1958) / 9–4–2003
Dorothy Kaufman (1950–1954) / 02–04–2003	Annie Cole (1956–1958) / 03–11–2003
Alice Web (1950–1954) / 10–02–2003	Norman Good (1956–1959) / 03–18–2003
Sandy Eycke (1950–1954) / 03–18–2003	Sherry Parson (1958–1962) / 03–07–2003
Robert Heart (1952–1956) / 6–24–2003	Nat Right (1956–1963) / 07–19–2004
Lou Anne Kensington (1953–1957) / 06–12–2003	Dotty Morris (1961–1965) / 09–24–2003
Nora Jones (1955–1959) / 06–12–2003	Burt Sadden (1963–1967) / 07–26–2004
	Betty Land (1964–1968) / 02–08–2003
	Linda Moss (1965–1969) / 09–24–2003
	Bud Land (1965–1969) / 08–16–2003
Border-Crossing Generation (1970–1985)	**Redivided Generation (1986–2000)**
Joanne Pet (1970–1974) / 10–16–2003	Harry Rice (1983–1987) / 02–25–2003
Sam Garnes (1970–1974) / 09–16–2003	Roberta Jones (1987–1991) / 09–02–2003
Pat Baley (1971–1975) / 10–20–2003	Cherry Gate (1988–1992) / 10–23–2003
David Randle (1972–1976) / 11–07–2003	Heather Korn (1989–1993) / 02–25–2003
Josh White (1972–1976) / 10–01–2003	Sue Cohen (1991–1995) / 09–2003
Michael Hallner (1972–1976) / 9–2003	Cecilia Hood (1992–1996) / 10–7–2003
Jeremy Garnes (1974–1978) / 09–09–2003	Betty Ames (1995–1999) / 08–19–2003
Timmy Whittle (1977–1981) / 06–28–2004	Bill Jackman (1995–1999) / 10–18–2003
Teresa Randle (1977–1981) / 06–09–2004	Vera Debin (1995–1999) / 01–15–2004
Sophie Baker (1981–1985) / 06–17–2004	Ivan Strasky (1998–2002) / 11–06–2004

Note: Interview with teachers was conducted on 11/07/2003: Michelle Shaw (English)—began teaching at Miller High in the 1970s; Julia Mills (coach)—began teaching at Miller High in the 1960s; Sam Buck (Spanish)—began teaching in the 1970s.

graduation, or colleges they attended. For example, if a narrator testified to having worked after school "to survive" and reported taking a job at the local factory after graduation, I deduced that the narrator broadly belonged to the working class at the time he attended high school. See

Table A.2 Distribution by race, gender, and class (at the time of school attendance) of graduates across generations. "NI" indicates that graduates did not identify themselves by class. Class identifications are reported in language used by graduates

Divided Generation (1950–1969)	Divided Generation (1950–1969)
Before Desegragation (1950–1969)	**After Desegregation (1956–1969)**
—1 white woman / poor	—3 African American women / middle class: 1 lived in an upper-middle-class house in a richer neighborhood, and 1 worked after school to help family financially.
—6 white women / middle class: 3 of them were college-bound, suggesting greater financial wealth than the other 2 who stated not having been able to afford college.	—1 African American man / rich
—1 white man / poor	—1 African American man / middle class
	—1 African American man / NI: he had to work after school "to survive" suggesting perhaps lower middle class / working class
All white graduates	—2 white women / middle class: both were college-bound and part of the "doers" identified by later generations as the "preps," suggesting the higher end of middle class
7 women	—1 white woman / middle class: but, she could not afford to go to college
1 man	—1 white man / middle class: he was going to college and his father owned a lucrative business suggesting higher end of middle class
	6 African American graduates
	4 white graduates
	6 women
	4 men

Border-Crossing Generation (1970–1985)

—2 African American women / middle class
—2 African American men / NI: their testimonies about where they lived when attending school suggest the broad category of middle class
—5 white men / NI: testimonies of one of them suggested that his background was working class and had worked in a factory after high school; I know where the other graduates lived when attending high school suggesting a middle-class background. The father of one of them was a small business owner.
—1 white woman / middle class

4 African American graduates
6 white graduates
3 women
7 men

Redivided Generation (1986–2000)

—2 white men / middle class
—4 white women / middle class: 1 was college-bound while the other 3 were very sensitive about not having the consumer power that the "preps" held; one of them in particular identified as a "farm girl" and her husband had been a work-release student at Miller High who later became a mechanic.
—2 women / middle class: preps, suggesting upper end of middle class
—1 white Russian immigrant man / NI: the house where he grew up and in which I interviewed him was upper-middle class (neighborhood where houses sold for 500,000.00 dollars in 2003)
—1 white Russian immigrant woman / middle class

0 African American graduates
10 white graduates, of which 2 Russian Immigrants
7 women
3 men

Note: One African American and one white woman, graduates of 1972 and 1985, spontaneously shared during their interviews that everyone at Miller High was middle class.

Table 2 in this section for a summary of narrators' class identifications by gender and race.

Transcribing and Coding

As much as possible, I transcribed interviews as soon as possible. Each transcription yielded on average twelve single-spaced pages. When vocal inflections or laughter qualified a remark in a striking way, I made note of it in the transcript. I spent an average of seven hours of transcribing per one hour of recording. I proceeded with analysis alongside data collection as I transcribed. The very act of transcribing was for me, in many ways, an important part of analysis. It was during the transcribing process that I jotted notes about comments I had not probed to my satisfaction, to revisit with the alumnus. Transcribing became more than just a preliminary stage to coding and analysis; it was the first stage of analysis as graduates' voices began echoing or contradicting each other. Transcribing kept the voices loud and alive in my mind.

I coded manually, without the use of a software program;[16] and I coded chronologically, beginning with the transcripts of those alumni who graduated in the 1950s, then the 1960s, and so on. Within each decade, I first created profiles for each alumnus, after which I looked for patterns across graduates' experiences. I did not a priori assume that being black or white or female or male, or other, would per force create similar experiences within those categories. It was important that I allow identifications to emerge from students' remembered lives, within their time and place, as they may. In the process, I developed my own idiosyncratic coding style.

For each transcript I created accompanying coding sheets on which I identified the alumnus, and where I organized themes that I had identified for the particular transcript. All of my coding sheets were written by hand, in pencil, allowing me to easily rearrange themes; these were attached to transcripts. Coding sheets were first completed for each individual transcription and then compared across transcripts within decades. Transcripts that shared most themes and shared language/linguistic register were grouped. From there, my analysis proceeded in narrative form. I compared experiences between groups within decades and then across decades. At this stage the narratives were two or three pages long per group and mostly written in syncopated prose. It is during this last stage of comparison (across decades) that generations emerged as groups consolidated.

As discussed in the Introduction, the echoed memories of student relationships as reported by those graduates interviewed across race, class, gender, and other divides, as well as memories collected through informal interviews over a two-year period (2003–2005), suggested that enough of those Miller High graduates interviewed perceived cross-group relationships similarly within certain time periods, as well as differently across time periods, to constitute generations; and that enough of them articulated in similar language those similarities out of which emerged time periods, which were further corroborated by broader demographic changes in Miller Town. Bound by shared memories of institutional practices, shared demographic configurations, and shared articulations (in use of vocabulary and cultural references) of cross-group relationships, the oral historical testimonies sketched the perimeters of generations.

It is also important to underscore that when certain voices were missing—for example, African American voices of the 1990s—their presence came into view obliquely through the testimonies of those interviewed, and were further brought out of the shadows, as much as possible, through yearbook and other documentary data and conversations with Miller High residents, black and white, over a two-year period (2003–2005). Sadly, Latino, Asian-Indian, and other voices of the 1990s are missing completely—the focus of further research.

Memory-Elicited Data

Within the world of memory, experience has been interpreted, perceptions have been sifted and settled, and meanings have been attributed. Within memory lingers the lasting impression.[17] Evidential issues surrounding the question of memory, while sensitive, held less weight in this analysis than they might have had I attempted to investigate, for example, the reputation of a particular Miller High teacher or administrator through students' remembered oral testimonies; or had I sought to construct an overarching collective memory of a homogeneous group of people with shared interests in representing their reality unilaterally.

Still, one might critique use of recollections as a means to capture students' interpretations since the interpretation of one's experience while experiencing it may have changed as one gained maturity. What circumvents this problem, in this study, is the live coproduction of the interview, which allows for a distinction in the present context, during the interview making, between the interpreted experience at the time of the memory making, and the reinterpreted experience a posteriori. For example, one alumnus recalled thinking that homework was useless, but today thinks that homework in general might have some grounding value. Of interest to me, within the context of this research, and regardless of the alumnus's present-day reinterpreted evaluation of homework, is that when he was a student, he thought homework was useless. Sensitive probing usually reveals the weight of present interpretations of past experience, and informants are usually interested, and in some cases even relieved, to explore those differences.

In this sense, the making of an interview allows for analytic distinctions that cannot be as readily made when mining such sources as diaries, journals, or biographies, where a posteriori reinterpretations are more easily slipped into the narratives by those who lived the history and are not as readily evident to the researcher. Moreover, it is in the comparing of experiences across situated perspectives that I gauged to what extent one graduate's understanding, at the time of his/her school attendance, might have been an idiosyncratic understanding and to what extent it was shared and with whom.

Archival Research

Archival research was conducted in the Baltimore County Board of Education archives in Towson, Maryland, where general county school policies were examined; in the Maryland rooms at Enoch Pratt Baltimore City Public Library and the University of Maryland College Park, where Maryland State Board of Education communications and census data were examined; in the library of Miller High School, where the school newspaper and archived alumni achievements and photographs were examined; and the "historic room" of Miller Town community library, where Miller High yearbooks and local community newspapers were examined. In the Miller Town library I also referenced histories of peoples of Baltimore County.

Locating records of student populations proved impossible. Miller High does not keep information older than five years (when I visited, the school had undergone a recent "major cleanup," as the administrative assistant explained), and the information available at the board of education tracked population shifts only by geographic sections, not by individual schools. Therefore, I tabulated student populations by counting senior pictures in yearbooks, which I compared against county and Miller Town census data. Moreover, various additions to the 1930s building, documented in the *Alumni's Directory*, further confirmed testimonies of changes in population at Miller High. Finally, I acquired a good sense of population changes through editorial comments expressed in the *Community Times* over time, as well as through my own encounters with the people of Miller Town between 2003 and 2005.

Of particular interest to this study is the use I made of yearbooks, inspired by the work of youth historian Paula Fass, who, in her essay "Creating New Identities: Youth and Ethnicity in New York City High Schools in the 1930s and 1940s,"[18] opened up new ways to consider yearbook data. Fass analyzed the distribution of students by ethnicity in various extracurricular activities. Following her lead, I developed a systematic record keeping of the distribution of students by gender, race, and ethnicity in the various extracurricular activities from yearbooks for every other year beginning with 1954.

For example, when averaging student participation by gender and race for the student "yearbook club," "newspaper club," and "honor society" (for yearbooks dated 1954, 1956, 1958, and 1960, 1962, 1964, 1966, 1968), the following proportions emerged: on average, thirteen white female students compared to eight white male students served on the yearbook staff, and no black students served across the two decades; on average, twelve white female students compared to five white male students served on the school newspaper staff, and no black youth (only one anomaly appears in 1964, when there were only ten white female students compared to fourteen white male students); on average, 126 young female students compared to seventy-eight white male students, and no black students, were in the honor society across the decades. Future Teachers of America and Future Nurses of America were consistently dominated by white women (only one white male appears in the late 1960s in Future Teachers of America, and two African American females appear in Future Nurses of America in the middle and end of the 1960s). Across the two decades, only women are represented in the Future Business Leaders of America across races, the extracurricular activity that recruited young women from the commercial track preparing to become secretaries. Populated solely by males were the Future Farmers of America club and, in the late 1960s, the "chess club." Also, young black men were enrolled in basketball and/or track, and of those most were also attending academic tracks (this could be easily corroborated with senior graduating pictures that included students' academic and extracurricular credentials). Football would not show up until 1969. There are no sports pictures featuring female African American students between 1950 and 1969. They do appear in the 1970s and thereafter. Such analyses were conducted for all the decades. Also, see Table A.3 at the end of this section for an organized tabulation of Miller Town population growth derived from census data.

I would like to end this section by reiterating the relevance of intersectionality theory in oral historical research that explores generational

Table A.3 Population Growth in Miller Town

	Total	White	Total Population	
			Identified as Other than white	Identified as Other than white ~ %
1950	2,077	1,912	165	~ 8%
1970	14,037	13,724	313	~ 2%
1980	19,385	16,928	2457	~ 13%
2000	22, 438	16,467 **Russian Population: (1413) 6.3%**	5971	~ 28% Black or African American: 18.4% Hispanic or Latino of any Race: 4.4% Asian: 4% Other: 1.6%

Source: Census Data / Baltimore City Enoch Pratt Free Public Library, *Maryland Room*
U.S. Department of Commerce Bureau of the Census 1950, Vol.1, Chapter 20
U.S. Department of Commerce Bureau of the Census 1970, Vol. 1, Chapter 20
U.S. Department of Commerce Bureau of the Census 1980, Microfiche / Tape File 3A / Place: <Miller Town>

transformations in the relational experiences of youth attending desegregated schools in the latter part of the twentieth century. I suggest that theory lurks in any intellectual inquiry—acknowledged, its influence can be argued; and the historian lurks in any historical inquiry—acknowledged, intellectual authority can be redistributed.[19]

Notes

Preface

1. See James Anderson and Dara Byrne, eds., *The Unfinished Agenda of Brown v. Board of Education* (Hoboken, NJ: John Wiley & Sons, 2004), for discussions regarding the legacy of *Brown*.
2. For a discussion of school desegregation policies and competing arguments regarding *Brown v Board* versus the *Coleman* report, see Amy Stuart Wells et al., "How Society Failed School Desegregation Policy: Looking Past the Schools to Understand Them," *Review of Research in Education*, 28 (2005): 47–99.
3. Wayne Urban and Jennings Wagoner, *American Education* (New York: McGraw Hill Companies, Inc., 2000), 288–289. Urban and Wagoner suggest that *Brown v. Board* "may have been one of the few occasions in our history when an educational policy was the catalyst for substantial changes in social relations and policies outside of school." Certainly, a series of legislations followed during the 1960s to enforce the resisted implementation of *Brown v. Board*—legislations such as the *Civil Rights* Act of 1964 that would indeed affect social relations outside of school. However, as some historians have argued, the political and judicial changes that ensued during the civil rights movement following *Brown v. Board* cannot be easily or primarily attributed to *Brown v. Board of Education* and must take into account the pre-*Brown* struggles that laid the foundation for the civil rights movements. For example, see John Dittmer, *Local People: The Struggle for Civil Rights in Mississippi* (Champaign, IL: University of Illinois Press, 1995). See also Jack Dougherty, "From Anecdote to Analysis: Oral Interviews and New Scholarship in Educational History," *Journal of American History* 86 (September 1999): 712–723, for reticence by many black communities to embrace school desegregation. For resistance to desegregation in the north, see the classic work by Anthony Lukas, *Common Ground: A Turbulent Decade in the Lives of Three American Families* (New York: Vintage Books, 1985).
4. "The 1965 Amendments to the Immigration and Nationality Act: repealed national origins restrictions which restricted immigration visas according to

the ethnic composition of the 1920 U.S. population. Before 1920 the United States had a very high rate of immigration. The 1965 amendment essentially returned immigration policy to the pre-1920 policy; and the 1990 Immigration Act: permitted the entry of 150,000 more legal immigrants annually"; http://sorrel.humboldt.edu/~economic/econ104/immigrat/ (January 2006). See also Alejandro Portes and Rubén G. Rumbaut, *Immigrant America: A Portrait* (Berkeley and Los Angeles: University of California Press, 1996).

5. For an in-depth discussion of multiculturalism as it has shaped education in the United States, see James Banks and Cherry McGee Banks, eds., *Handbook of Research on Multicultural Education* 2nd ed. (San Francisco: John Wiley & Sons, 2004).

6. See Caroline Eick and Linda Valli, "Teachers as Cultural Mediators: A Comparison of the Accountability Era to the Assimilation Era," *Critical Inquiry in Language Studies* 7 (September 2010).

7. Refer to Anderson and Byrne, *The Unfinished Agenda of Brown v. Board of Education*, earlier cited.

8. Arthur B. Kennickell, "A Rolling Tide: Changes in the Distribution of Wealth in the U.S., 1989–2001," Levy Economics Institute, Table 10 (November 2003), http://www.faireconomy.org/research/wealth_charts.html (January, 2006).

9. Gary Orfield, "*Schools More Separate: Consequences of a Decade of Resegregation,*" Harvard University, Civil Rights Project website, http://www.civilrightsproject.harvard.edu/research/deseg/separate_schools01.php (January 2006). See also Peter Irons, *Jim Crow's Children: The Broken Promise of the Brown Decision* (New York: Viking, published by Penguin Group, 2002).

10. See U.S. Hate Crimes: Definitions and Facts, http://www.religioustolerance.org/hom_hat3.htm (January 2006).

11. Although the military has been considered, especially with the recent inclusion of women, as our most integrated institution since World War II, it is, except for drafts (which ended with the military withdrawal from Vietnam in 1973), an all-voluntary military.

12. See Patrick J. Ryan, "A Case Study in the Cultural Origins of a Superpower: Liberal Individualism, American Nationalism, and the Rise of High School Life, A Study of Cleveland's Central and East Technical Schools, 1890–1918," *History of Education Quarterly* 45 (Spring 2005): 66–95.

13. Ibid.

14. "Massification" refers to the expansion of mass education as part of modernity. See John Boli, Francisco O. Ramirez, and John W. Meyer, "Explaining the Origins of Expansion of Mass Education," *Comparative Education Review* 29 (1985): 145–170. Also, Marginson explains how educational aims will increasingly focus on developing capacities for adaptation to new environments and foreign associations. He contrasts this latest development to education of the Enlightenment, which emphasized strong links between family, work, and citizenship. See Simon Marginson, "After Globalization, Emerging Politics of Education," *Journal of Education Policy* 14 (1999): 19–35.

15. Benjamin Barber quoted in R.C. Salomone, *Vision of Schooling: Conscience, Community, and Common Education* (New Haven: Yale, 2000), 11.
16. Howard Zinn, *A People's History of the United States* (New York: Harper Collins, 2003), 11.
17. Dynamics of acculturation and minority students' many responses to forces of assimilation, from resistance to accommodation, have been widely studied by educational ethnographers. For example, see works by Michelle Fine, *Framing Dropouts* (Albany: State University of New York Press, 1991); Anne Locke Davidson, *Making and Molding Identity in Schools* (New York: State University of New York Press, 1996); Jeremy Price, *Against the Odds* (Stamford, CT: Alex Publishing Corporation, 2000); by anthropologists, including John Ogbu, "Immigrant and Involuntary Minorities: A Cultural-Ecological Theory of School Performance with Some Implications for Education," *Anthropology and Education Quarterly* 29 (1998): 155–188; and by historians of education such as V.P. Franklin, "Black High School Activism in the Late 1960s: An Urban Phenomenon?" *Journal of Research in Education* 10 (2000): 3–8; Victoria-María MacDonald, *Latino Education in the U.S.: A Narrative History, 1531–2000* (2004).
18. Term used by Marginson. See Simon Marginson, "After Globalization, Emerging Politics of Education," earlier cited.
19. Howard Zinn, *A People's History of the United States*, 10.
20. Amy Gutmann, *Democratic Education*, 309.
21. Maxine Green quoted by Judith A. Ramaley, The Presidents/Fourth of July Declaration on the Civic Responsibility of Higher Education, http://www.compact.org/resources/plc-declaration.html (December 2005).

Introduction

1. I borrow from Fine, Weiss, and Powell to define integration as "intellectual and social engagement across racial and ethnic groups." See Michelle Fine, Lois Weiss, and Linda C. Powell, *Harvard Educational Review* 67 (Summer 1997): 248. Thus I contrast the notion of *integration* to that of *desegregation*, which refers foremost to admission policies and processes to secure laws that ensure equal representation of students across race in schools.
2. Amy Gutmann, *Democratic Education* (Princeton, NJ: Princeton University Press, 1999), 161. See also Amy Stuart Wells et al., "How Society Failed School Desegregation Policy: Looking Past the Schools to Understand Them," *Review of Research in Education* 28 (2005): 47–99. Wells and her coauthors report how "recorded attitudes had changed relatively dramatically in the 40 years since the Brown decision...the percentage of Americans of all races who believed that the Supreme Court was right in its *Brown* decision increased from 63% in the early 1960s to 87% in the mid-1990s...percentage of Americans who believed that more should be done to integrate schools had risen rapidly in short time span, from 37% in 1988 to 56% in 1994," p. 77.

The authors use the word "integration" rather than "desegregation" to refer to balanced representation of students across racial groups.
3. See Jeanie Oakes, *Keeping Track: How Schools Structure Inequality* (Yale University Press, 2006).
4. Schofield (1989) quoted in Amy Stuart et al., "How Society Failed School Desegregation Policy: Looking Past the Schools to Understand Them," *Review of Research in Education* 28 (2005): 71.
5. Jencks quoted in Amy Stuart Wells et al., "How Society Failed School Desegregation Policy: Looking Past the Schools to Understand Them," *Review of Research in Education* 28 (2005): 71.
6. Gerald Grant, *The World We Created at Hamilton High* (Cambridge, MA: Harvard University Press, 1988). This analysis also builds closely on the work of William Graebner, *Coming of Age in Buffalo: Youth and Authority in the Postwar Era* (Philadelphia: Temple University Press, 1990). With the works of Grant and Graebner, this history shares its focus on youth intergroup relations across race divides (Grant) and across race and class and gender divides (Graebner). Unlike Graebner's work, but similar to Grant's work, it situates its youth within school boundaries. However, while both Grant and Graebner examine youth in cities (Grant's city counted 220,000 people in 1953 compared to Miller Town's 2,077 in 1950—and Graebner's city was Buffalo), this analysis focuses on a third-tier suburban town that develops over time from rural-suburban to urban-suburban, and it examines more closely how peer relationships are historically shaped across markers of difference within the extracurricular and curricular spaces of school. More important, it identifies students' situated positions within intersections of gender, class, race, religion, and nationality. To my knowledge this is the only other history of education that examines youth intergroup relationships, and the only history of education that examines young people's experiences from multiple intersecting categories of identity. In general, cultural historians and historians of youth have submerged students' experiences under the weight of economic, demographic, and political analyses. They have preferred to situate young people outside the perimeters of high school, and have represented them one-dimensionally. See John Gillis, *Youth and History: Tradition and Change in European Age Relations 1779–Present* (NY and London: Academic Press, 1974); Joseph Kett, *Rites of Passage: Adolescence in America, 1790 to the Present* (NY: Basic Books, 1977); Stephen Shlossman, *Love and the American Delinquent* (Chicago: University of Chicago Press, 1977); John Gilbert, *Cycles of Outrage: America's Reaction to the Juvenile Delinquent in the 1950s* (NY: Oxford University Press, 1986); R. Cohen, "The Delinquents: Censorship and Youth Culture in Recent U.S. History," *History of Education Quarterly* 37 (1997): 251–270; John Modell, *Into One's Own: From Youth to Adulthood in the United States, 1920–1975* (Berkeley: University of California Press, 1989); Grace Palladino, *Teenagers: An American History* (Chicago: University of Chicago Press, 1996); Joe Austin and Michael Willard, eds., *Generations of Youth: Youth Culture and History in Twentieth-Century America* (NY: New York University Press, 1998). The few histories that do situate youth within the high school setting and that attempt

to address gender, class, and racial differences do not reveal interactions between students across gender, race, and class boundaries. See E.F. Frazier, *Negro Youth at the Crossroads: Their Personality Development in the Middle States* (NY: Scholar Books, 1967); Kenneth Fish, *Conflict and Dissent in the High School* (NY: Bruce Publishing Co., 1970); Paula Fass, "Creating New Identities: Youth and Ethnicity in New York City High Schools in the 1930s and 1940s," in *Generations of Youth: Youth Culture and History in Twentieth Century America*, Joe Austin and Michael Willard, eds. (NY: New York Press, 1998), 95–117.

7. See addendum "Methodology: The Transparent Historian" for an in-depth discussion of oral histories as social acts and the historical consciousness.

8. Most recent works on high school activism include Dwayne C. Wright, "Black Pride Day, 1968: High School Student Activism in York, Pennsylvania," *The Journal of African-American History* 88 (Spring 2003): 151–162; Dionne Danns, "Chicago High School Students' Movement for Quality Education, 1966–1971," *The Journal of African-American History* 88 (Spring 2003): 138–150. See also V.P. Franklin, "Black High School Student Activism in the 1960s: An Urban Phenomenon?" *Journal of Research in Education* 10 (Fall 2000): 3–8. For an overview of works on African-American student activism in colleges as well as secondary institutions, see V.P. Franklin, "Introduction: African American Student Activism in the 20th Century," *The Journal of African-American History* 88 (Spring 2003): 105–109.

9. See Robert Palumbos, "Student Involvement in the Baltimore Civil Rights Movement, 1953–63," *Maryland Historical Magazine* 94 (1999): 449 and 485.

10. Information gathered from discussions with Howell S. Baum, who is presently working on a book entitled *How Liberalism Failed Brown: School Desegregation in Baltimore*. Howell Baum is professor of urban studies and planning at the University of Maryland, College Park. See also Steven Hahn, *A Nation Under Feet: Black Political Struggles in the Rural South from Slavery to the Great Migration* (Harvard University Press, 2004). In this award-winning history, Hahn brings to light the role of biracial alliances in African Americans' quest for self-governance, alliances that echo the African American and white alliances addressed in this history, whether in Baltimore prior to desegregation or in Miller Town following desegregation.

11. Thomas J. Vicino, *Transforming Race and Class in Suburbia: Decline in Metropolitan Baltimore* (NY: Palgrave Macmillan, 2008), 42. Between 1950 and 1970, the total population of Miller Town rose from 2,077 to 14,037. In 1950, whites represented approximately 92 percent of the population, and blacks were roughly 8 percent. In 1970, as white flight intensified, the percentage of whites rose to 97 percent of the population (from 1,912 registered white citizens in 1950 to 13,724 in 1970) compared to the black population, which, while it decreased in proportion to approximately 3 percent of the total population, increased from 165 registered African Americans in 1950 to 313 in 1970. See Table 3 for census data from the U.S. Department of Commerce Bureau of the Census.

12. See works by Louis S. Diggs: *Since the Beginning: African American Communities in Towson* (Baltimore: Uptown Press, 2000); *Holding on to Their Heritage* (Catonsville, MD: privately printed, 1996); *In Our Own Voices: A Folk History in Legacy* (Catonsville, MD: privately printed, 1998).
13. For discussions of changing immigration patterns since the Immigration Act of 1965 and implications for school populations, see works by Laurie Olsen, *Made in America: Immigrant Students in Our Public Schools.* (NY: New Press, 1997); and Alejandro Portes and Rubén Rumbaut, *Immigrant America: A Portrait*, 2nd ed. (Berkeley: University of California Press, 1996).
14. Gerald Grant, *The World We Created at Hamilton High*, 6.
15. For elaborations on multivocal narratives, see Peter Burke, *History and Social Theory* (NY: Cornell University Press, 2005). For ethnographies that examine students' high school experiences across markers of difference, see August Hollingstead, *Elmtown's Youth: The Impact of Social Class and Adolescents* (NY: J. Wiley, 1949); Penelope Eckert, *Jocks & Burnouts: Social Categories and Identity in the High School* (NY: Teachers College Press, 1989); Michelle Fine, *Framing Dropouts* (Albany: State University of New York Press, 1991); Ellen Brantlinger, *The Politics of Class in Secondary School* (NY: Teachers College Press, 1993); Jay MacLeod, *Ain't No Makin' It* (Boulder, CO: Westview Press, 1995); Ann Locke-Davidson, *Making and Molding Identity in Schools* (NY: State University of New York Press, 1996); Laurie Olsen, *Made in America: Immigrant Students in Our Public Schools* (NY: New Press, 1997).
16. See Patricia Collins, *Black Feminist Thought: Knowledge, Consciousness, and the Politics of Empowerment* (NY: Routledge, 2000); Patricia Collins, *Fighting Words: Black Women and the Search for Justice* (Minneapolis: University of Minnesota Press, 1998).
17. See Kimberlé Crenshaw, Neil Gotanda, Gary Peller, and Kendall Thomas, eds., *Critical Race Theory: The Key Writings that Formed the Movement*, (NY: New York Press, 1995); Bill Ashcroft, Gareth Griffiths, and Helen Tiffin, *Post-Colonial Studies: The Key Concepts* (NY: Routledge, 2005); as well as the works of Patricia Collins earlier cited.
18. Leslie McCall, "The Complexity of Intersectionality," in *Signs: Journal of Women in Culture and Society* 30 (2005), http://www.journals.uchicago.edu/doi/abs/10.1086/426800.
19. Alessandro Portelli, "What Makes Oral History Different," in *The Oral History Reader*, ed. Robert Perks and Alistair Thomson (NY: Routledge, 2005), 63–74 (first emphasis mine, second emphasis that of the author). See also Alessandro Portelli, *The Death of Luigi Trastulli and Other Stories: Form and Meaning in Oral History* (Albany: SUNY Press, 1990).
20. Thirty-seven alumni of different races and socioeconomic backgrounds, identified through yearbooks and alumni directories as well as through snowball sampling, and who graduated from Miller High between 1954 and 2002, were formally interviewed in tape-recorded sessions that ranged from one to two hours, and informally through follow-up calls, visits, and e-mail communications. These alumni were chosen for their differences (gender and class just prior to 1956 desegregation; gender, class, and race following

21. In examining yearbooks, total populations of students and distribution of student population by race and gender were tabulated, as well as distribution of students across race and gender in extracurricular activities and clubs. For descriptions of archival research, see Addendum: Methodology.
desegregation), and situated as males and females of different racial and economic backgrounds, across the decades, as much as possible. The decade boundaries emerged from changes in oral testimonies as well as corroborating changes examined in yearbooks, community newspapers, census data, and other documentary sources. See Addendum: Methodology.
21. In examining yearbooks, total populations of students and distribution of student population by race and gender were tabulated, as well as distribution of students across race and gender in extracurricular activities and clubs. For descriptions of archival research, see Addendum: Methodology.
22. Irma Omeldo, "Redefining Culture through the Memorias of Elderly Latinas," *Qualitative Inquiry* 5 (1993): 353–376.
23. See Roger Brown, *Social Psychology* (NY: Free Press, 1986); see also Robert Levine and Donald Campbell, *Ethnocentrism: Theories of Conflict, Ethnic Attitudes, and Group Behavior* (NY: John Wiley & Sons, 1972).
24. See Roger, J.R. Levesque, *Dangerous Adolescents, Model Adolescents: Shaping the Role and Promise of Education* (NY: Kluwer Academic/Plenum Publishers, 2002), 53–54. In discussing the ruling in *Tinker vs. Des Moines Independent Community School District*, 1969, Levesque explains, "[S]tudents challenged a school's prohibition against wearing black arm bands in protest of the Vietnam War." Students won as it was judged that "minors were persons protected by the Constitution."
25. Zero-tolerance policies, which spread in the mid-1990s, targeted possession or use of drugs or alcohol and weapons by students on school premises. In some cases, zero-tolerance policies became opportunities for school authorities to threaten youth beyond evidence of wrongdoing. For opinions in support of zero-tolerance policies, see National Safety Guard and Security Services, Zero Tolerance, http:www.schoolsecurity.org/trends/zero_tolerance.html. For critiques, see Dennis Cauchon, "Zero Tolerance Policies' Lack of Flexibility," *USA Today*, April 13, 1999.
26. See Tyack and Cuban, *Tinkering Toward Utopia: A Century of Public School Reform* (Cambridge, MA: Harvard University Press, 1995), 44–45. Tyack and Cuban remark about the politically conservative periods of the 1950s and 1980s: "Policy talk about schools stressed a struggle for national survival and international competition...In such periods, policy elites want to challenge the talented, stress the academic basics and press for greater coherence and discipline in education." The authors suggest that "liberal eras such as the 1930s and 1960s stress ideology of access and equality." Tyack and Cuban remark, however, that by the close of the twentieth century, in the "late 80s and 90s...conservatives and liberals alike...called for national standards." By the end of the twentieth century, cyclical, pendulum-like criticisms of high school performance reach the highest level of attack on the legitimacy of the high school within the 1950–2000 period. See also work by William Wraga, *Democracy's High School: The Comprehensive High School and Educational Reform in the United States* (Lanham, MD: University Press of America, 1994).
27. See Grace Palladino, *Teenagers: An American History* (NY: Basic Books, 1996).

28. See report by Gary Orfield and Chungmei Lee, "Brown at Fifty: King's Dream or Plessy's Nightmare," Civil Rights Project Harvard University, http://www.civilrightsproject.ucla.edu/research/reseg04/resegregation04.php (accessed June 24, 2010).
29. See Clayborne Carson, "Two Cheers for Brown v. Board of Education," *Journal of American History* 91 (June 2004), 26–31.
30. Carnoy and Levin, quoted in David F. Labaree, *How to Succeed in School Without Really Learning* (New Haven: Yale University Press, 1997), 49.
31. Wells et al., "How Society Failed School Desegregation Policy: Looking Past the Schools to Understand Them," *Review of Research in Education*, 28 (2005): 90; "[C]lose friendships within racially diverse schools are more often than not same-race friendships...." This historical analysis shows, however, that interracial friendships can flourish under particular social and demographic conditions such as those experienced by the Border-Crossing Generation.
32. Findings by Amy Stuart Wells et al. also suggest "that there are several ways to interpret lunchroom segregation within desegregated schools... and one is to consider who students see and interact with outside of school" (Wells, 2005, p. 90). Certainly there are indications that even at the beginning of desegregation at Miller High, during the Divided Generation interracial playmates of Miller Town were more likely to sit together in the cafeteria during lunchtime. Refer to Annie Cole's testimonies. I suggest that beyond imported habits of association, students also segregated in the cafeteria at the end of the century because of the absence of adult role-modeling and adult engagement.
33. For a discussion of the nefarious role of tracking on student achievement, see Jeanie Oakes, *Keeping Track: How Schools Structure Inequality* (Yale University Press, 2006).

1 Memories of Class, Race, and Gender Divisions: Immediate Pre- and Post-Desegregation Years (1950–1969)

1. The name of the school and all people's names have been replaced with pseudonyms to protect those graduates unwilling to be identified by testimonies they felt might compromise their professional and familial relationships in the community. Some narrators have agreed to reconsider release of transcripts at a later date. See Addendum: Methodology.
2. <Miller> High was first known as <Miller> Academy, founded in 1820. It was renamed <Miller Town> High School in the 1840s, and then renamed <Miller> High in the 1950s. For details about the various site and building transformations of the high school, I referred to C.P., *<Miller's> Century of Progress 1878–1978* (<Miller Town>: C.P., 1978), located in the <Miller Town> Library archives.

3. <Miller High> *Alumni Directory* (NY: Bernard C. Harris Publishing, 2000), ix. The 1930s building is further described as housing "the widest variety of educational equipment ever seen in this section, making it possible for the school to offer courses of instruction in practically every phase of academic, cultural, commercial, and industrial secondary education."
4. Sherry Parson (1958–1962), interview with the author, March 7, 2003.
5. C.P., <*Miller's*> *Century of Progress: 1878–1978* (<Miller Town>: C.P., 1978).
6. Sherry Parson (1958–1962), interview with the author, March 7, 2003. The bell from the original academy, built in the early nineteenth century, was preserved and kept in the high school throughout its transmutations. It is presently in a small garden and is still rung at the beginning of every school year.
7. From conversations with Annie Milligan (see Addendum: Methodology). See also Louis S. Diggs, *Since the Beginning: African American Communities in Towson* (Baltimore: Uptown Press, 2000); *Holding on to Their Heritage* (Catonsville, MD: privately printed, 1996); *In Our Own Voices: A Folk History in Legacy* (Catonsville, MD: privately printed, 1998). See also E. Franklin Frazier, *Black Bourgeoisie* (NY: Free Press Paperbacks, 1997). Frazier's work, originally published in 1967 and widely criticized for its harsh portrayal of middle-class blacks in the United States, problematizes the meaning of "middle class" for African Americans whose middle-class status did not de facto change their relative economic standing in the United States. Frazier's findings regarding conspicuous consumption among the African American middle class do not reflect the black community of Miller Town in this history. However, Frazier's seminal work provides a context for understanding the economic place of African American business owners, and certainly those of the 1950 to 1969 time period discussed in this chapter.
8. Linda Eisenman, *Higher Education for Women in Postwar America, 1945–1965: Reclaiming the Incidental Student* (Baltimore: Johns Hopkins University Press, 2006), 19. See also Kenneth. T. Jackson, *Crabgrass Frontier: The Suburbanization of the United States* (Oxford, NY, Toronto: Oxford University Press, 1985).
9. Linda Moss (1965–1969), interview with the author, September 24, 2003.
10. At that time, high school graduates could still be teachers. Kaufman (1950–1954), interview with the author, February 4, 2003: "I could get a job in teaching in the County without a degree."
11. Plays were performed regularly not only by students, but also by faculty and community members. Annual fairs were held where farmers showed livestock and sold produce. It was also an opportunity for the townspeople to attend free sporting matches, art shows, and live music.
12. To get a sense of the school climate at Miller High in 1956, I interviewed students who graduated in 1954 and 1956. These graduates' experiences covered the full first half of the 1950s.
13. Dorothy Kaufman (1950–1954), interview with the author, February 4, 2003.

14. Dorothy Kauffman (1950–1954), interview with author, February 4, 2003.
15. Yearbooks consistently reveal a majority of white women in honor society photographs. See Addendum: Methodology.
16. Robert Heart (1952–1956), interview with the author, June 24, 2006: "They expected less of the boys... [the teachers] weren't going to get anything out of [the boys]." Judy Law (1950–1954), interview with the author, February 25, 2003: "Girls were considered to be smarter." Alice Web (1950–1954), interview with the author, October 2, 2003: "Girls were the better students." See also James B. Conant, *The American High School Today: A First Report to Interested Citizens* (NY: McGraw-Hill, 1959), and his second report, *The Comprehensive High School: A Second Report* (1967). Although there is no possibility of assessing how many girls did homework for the boys, the fact that male and female alumni who do not know each other reported as much, and that they reported as much across time periods (within the broader study) justifies documenting this recollection. Future research might focus on recovering changes and continuities in female-male student relationships not in terms of achievement gaps, but in terms of study habits and the role that sexual interest plays (or doesn't play) in shaping them.
17. Robert Heart (1952–1956), interview with the author, June 24, 2006. Tim Whittle, alumnus who attended in the late 1970s, also recalled girls doing boys' homework (see part II, chapter 3).
18. Alice Web (1950–1954), interview with the author, October 2, 2003.
19. Ibid.
20. Yearbooks and alumni stories reveal a consistent majority of young women yearbook editors, writers, and illustrators, as well as school newspaper editors and writers. See Addendum: Methodology.
21. Linda Moss (1965–1969), interview with the author, September 24, 2003.
22. Sandy Eycke (1950–1954), interview with the author, March 18, 2003: "A large portion of my class started first grade...went through all twelve grades...everybody went to Miller at some point in my family...You had teachers who had taught most of us. A lot of teachers lived in the community...A lot of them knew parents, brothers, and sisters." Betty Land (1964–1968), interview with the author, February 8, 2003: "My mom is a Miller High graduate, as is her mom and dad and parents. So I have a long line of Miller graduates in my family. I have two children of my own that have graduated from Miller High, and I have one there now and one in the middle school." Linda Moss (1965–1969), interview with the author, September 24, 2003: "They were role models...the relationships were great...and it was so relaxed but firm, and everybody did what they were supposed to do."
23. <*The Key*> yearbook (1958), 74.
24. Alice Web (1950–1954), interview with the author, October 2, 2003.
25. Ibid.
26. Sandy Eycke (1950–1954), interview with the author, March 18, 2003: "I probably would have gone to higher education if we could have afforded it. I think that was by default...if you weren't going to college, this is what

was left." Some, like Nora Jones, who was artistically inclined, completed their studies within the general track and availed themselves of an "absolutely wonderful art program...[and] great art teachers." Quote from transcript of audiotape interview with Nora Jones (1955–59), interview with the author, June 12, 2003.
27. Judy Law, interview with the author, February 25, 2003.
28. Judy Law (1950–1954), interview with the author, February 25, 2003. Also, Alice Web, ((1950–1954), interview with the author, October 2, 2003), in remembering Judy Law, said, "She didn't go to college...and she was one of the brightest in the class. She never thought she was going to go to college...we were, to a certain extent, in our slots."
29. Graduates' recollections echo historian Beth Bailey's findings regarding the social practice of dating and "going steady." See Beth Bailey, *From Front Porch to Back Seat: Courtship in Twentieth-Century America* (Baltimore: Johns Hopkins University Press, 1998).
30. Judy Law (1950–1954), interview with the author, February 25, 2003: "The biggest negative for me at high school is that it didn't really prepare me for the real world...it didn't prepare me to question authority."
31. Dorothy Kauffman (1950–1954), interview with the author, February 4, 2003.
32. Judy Law (1950–1954), interview with the author, February 25, 2003.
33. Nora Jones (1955–1959) and Lou-Anne Kensington (1953–1957), interview with the author, June 12, 2003.
34. Nat Right (1957–1963), interview with the author, July 19, 2004.
35. Bud Land (1964–1968), interview with the author, August 16, 2003.
36. Dorothy Kaufman (1950–1954), interview with the author, February 4, 2003.
37. Alice Web (1950–1954), interview with the author, October 2, 2003: "The English teachers that I had were really, really good, and social studies...but the math was just rote...math and science was not well taught...I would have been really interested in biology...I had Mr. B., whom everybody loved, but he was not a good teacher."
38. Judy Law (1950–1954), interview with the author, February 25, 2003: "For me it was a saving grace to be able to come to school. I got to dress up and came to school all pretty and clean in something my grandmother had made me...I traveled. I mean, it was travel from the country to the city...Oh, it was a big social time."
39. Judy Law (1950–1954), interview with the author, February 25, 2003: "[The diploma] meant that I had a job. And that I was free...to leave home. That was what it meant to me. My freedom. When they think of freedom, most people think of it as a patriotic thing. I think of it as a very personal thing."
40. Ibid. "It was a big deal that I graduated...I think it gave them an elevated level of prestige. My father wanted me to graduate because he never had; my mother wanted me to graduate because it was her school and because, I guess, it reflected on her in some way."

41. Dorothy Kaufman (1950–1954), interview with the author, February 4, 2003.
42. Alice Web (1950–1954), interview with the author, October 2, 2003.
43. Sherry Parson (1958–1962), interview with the author, March 7, 2003.
44. Linda Moss (1965–1969), interview with the author, September 24, 2003.
45. Dorothy Kaufman (1950–1954), interview with the author, February 4, 2003.
46. Ibid. Also, Alice Web (1950–1954), interview with the author, October 2, 2003: "Every time you had a class, she'd rearrange the class, and the person that got the best grade on the test would sit in the first seat in the first row. And then it would go all the way to the person that got the worst grade in the back of the room...I guess that shows you the advantage that the bright kids had, and very often the bright kids were also the ones who had more advantages at home, too. And got more support at home."
47. Betty Land (1964–1968), interview with the author, February 8, 2003: "Every Friday we had basketball, boys' basketball, and that was really a big thing then. And everybody would go to the games...Big school spirit. And we always had the cheerleaders. We always had a girl that was dressed up like...." Linda Moss (1965–1969), interview with the author, September 24, 2003: "We had great plays. We had great pep rallies."
48. Alice Weber (1950–1954), interview with the author, October 2, 2003: "And we all knew the whole story of graduation and we all just loved it, you know, the pomp and circumstance of it. We sang 'I'll never walk alone'...it was a tradition...I knew it was a place I was really tied to, it was part of my life."
49. Dorothy Kaufman (1950–1954), interview with the author, February 4, 2003: "I do remember one girl who...dropped out. Because her parents insisted that she [girl with learning disability] be allowed to come to school and to, you know, have the social experience. She was a sweetheart. But she was so lonely; you could tell she was lonely because she wouldn't come join in unless you went to get her. You know, she didn't have the confidence."
50. Alice Web (1950–1954), interview with the author, October 2, 2003: "There was one girl who was pregnant when we graduated, but she was wild, you know...everybody knew she was wild...She was like four or five months pregnant." Dorothy Kaufman (1950–1954), interview with the author, February 4, 2003: "One pregnant girl...just disappeared." Linda Moss (1965–1969), interview with the author, September 24, 2003: "There were pregnant girls in our class. Was pretty scandalous."
51. Alice Web (1950–1954), interview with the author, October 2, 2003.
52. Ibid.
53. Robert Heart (1952–1956), interview with the author, June 24, 2003.
54. Ibid.
55. Robert Heart's experiences as a poor farm boy echoes those of poor white sharecropping families recorded by the writer James Agee and the photographer Walker Evans in *Let Us Now Praise Famous Men: Three Tenant Families* (Boston: Houghton Mifflin/Mariner, 2001), originally published in 1941.

56. Robert Heart (1952–1956), interview with the author, June 24, 2003: "I would go on my paper route from one to three o'clock in the morning. Come home, grab a couple of hours sleep, then milk six cows from 5:30 to 6:30, then go to school. Get off school. Go home, milk six cows and then work for Shellborn in Miller Town at the gas station, from seven to eleven at night...I did that every day. And then, when the weekends came around, I would help on the weekend paper. That's what I did. That's how I got through high school...I was always sleepy....Most of the farm boys didn't have time to do their homework...They knew we didn't do it...We only got to learn what's in the class."
57. Robert Heart (1952–1956), interview with the author, June 24, 2003.
58. There was also a sexual edge to these boys for which, as "red blooded American boys," they expected to be understood. Robert Heart (1952–1956), interview with the author, June 24, 2003: "I got kicked out of one my English classes and I got sent to the principal's. And he asked me, 'What is going on in that classroom for you to be kicked out?' I said, 'All of the boys are sitting in the front of the classroom...she wanted us to sit in the front of the classroom. And she sits on the desk swinging her legs. Now, none of us could concentrate on anything.' So he walked down the hall. She was sitting on the desk with the boys in front. Well, after that I had no problems with her. I still got a D...They liked to pick on the farm boys...their dads weren't educated."
59. Robert Heart (1952–1956), interview with the author, June 24, 2003: "One of my favorite teachers was the shop teacher...He covered for us all the time...He got me through high school...Instead of spending two hours in detention, he would say, 'Come down here I need you. I need this work done. I need you to draw the floor plan.'...So I would go down to the shop class, sit at the drawing table, and draw...He got me pretty much through high school...He'd say, 'I'd love to grab you by the ear and say this is what you have to do to get out of here.' He helped me through a lot."
60. Ibid.
61. Ibid.
62. Annie Cole (1957–1959), interview with the author, March 11, 2003.
63. Robert Heart (1952–1956), interview with the author, June 24, 2003.
64. Ibid.
65. Sherry Parson (1958–1962), interview with the author, March 7, 2003.
66. Ibid.
67. Robert Heart shared how he would try mimicking the ways of those with money. He was acutely aware of being shunned by people whom he called "upper class," and he tried to date "rich" girls who continually refused him. It is also essential to point out that while farm boys across time periods were remembered as "hillbillies" and "rednecks" and reviled by alumni as racist, there was no evidence I could find in Robert Heart's recollections to suggest that he was a racist farm boy. It is equally important to note that Robert graduated in the summer of 1956, a couple of months before desegregation.
68. Ibid.

69. Robert Heart (1952–1956), interview with the author, June 24, 2003. Descriptions of high school as a prison-like or regimented place were echoed by other white lower-class male students across time periods, in particular Tim Whittle, a 1981 graduate (see part II, chapter 3).
70. Robert Heart (1952–1956), interview with the author, June 24, 2003.
71. Ibid.
72. Robert Heart (1952–1956), interview with the author, June 24, 2003: "If you made a mistake at the beginning of the first problem, you failed it...and then you would go back and do the problem, but he wouldn't do the problem for the speed of the slowest. He would do the problem for the speed of the fastest...So I failed his class...So I switched to Mrs. Hill, and she knew where I was coming from, okay [fatherless farm boy on welfare]...and then all of a sudden it makes sense...that's what she did and I got through it with no problems." This memory illustrates the allowances Robert expected precisely because of his life circumstances.
73. Robert Heart (1952–1956), interview with the author, June 24, 2003: "[We were] telling the teachers, look, we're going to school. Care about us. We may not be paying attention, but we are going to school."
74. Robert Heart (1952–1956), interview with the author, June 24, 2003.
75. Sherry Parson (1958–1962), interview with the author, March 7, 2003.
76. Robert Heart (1952–1956), interview with the author, June 24, 2003.
77. Bud Land (1964–1968), interview with the author, August 16, 2003: "I had had it [with high school]. Literature? I don't want to use it at all. I don't want to deal with it. Where is it going to fall into my life?"
78. Bud Land (1964–1968), interview with the author, August 16, 2003.
79. Ibid: "Betty went along with the very conforming group....You can look at the yearbooks and see Betty and her friends hanging around in ninth grade, and it's the same group in twelfth grade." Betty, as earlier discussed, was one of the white middle-class girls who understood themselves to be, as Linda Moss expressed, "the doers."
80. Bud Land (1964–1968), interview with the author, August 16, 2003.
81. Ibid.
82. Bud Land (1964–1968), interview with the author, August 16, 2003.
83. Doris Right (1956–1958), interview with the author, September 4, 2003.
84. Annie Cole (1956–1958), interview with the author, March 11, 2003.
85. Ibid.
86. Doris Right (1956–1958), interview with the author, September 4, 2003.
87. Annie Cole (1956–1958), interview with the author, March 11, 2003.
88. Ibid.
89. Ibid.
90. Doris Right (1956–1958), interview with the author, September 4, 2003.
91. Ibid: "My experiences with the teachers were great. As far as I knew them, I had no problems with the teachers, they appeared very helpful, especially Miss R., who I understand is deceased now. [Then there was] Miss T., was her name. She was my speech teacher...she made me talk...she was a sweet lady, she really was." Annie Cole (1956–57), interview with the author, March

11, 2003: "I was a good baseball player. And the gym teachers liked me a lot... Mrs. Streesby, she was a very nice teacher. She was one of the teachers that you could go and talk to. And she didn't take sides with anybody. She didn't like nobody to be called names, and you did get in trouble if you called names and she heard you. She was good like that... But it was a few of them."
92. Annie Cole (1956–1957), interview with the author, March 11, 2003.
93. Dorothy Kaufman (1950–1954), interview with the author, February 4, 2003.
94. Annie Cole (1956–1957), interview with the author, March 11, 2003.
95. Doris Right (1956–1958), interview with the author, September 4, 2003.
96. Annie Cole (1956–1957), interview with the author, March 11, 2003.
97. Doris Right (1956–1958), interview with the author, September 4, 2003.
98. Annie Cole (1956–1957), interview with the author, March 11, 2003 .
99. Ibid: "You know, the parents, the white parents had been there for years. Some of the families been in the community helping the school. It wasn't much he could do. He tried. He made things as comfortable as he could for us. I give him that."
100. Annie Cole (1956–57), interview with the author, March 11, 2003: "I'll tell you what made me really, really feel good. When I was sitting in my classroom one day and Mr. Lancaster called me over the intercom. Told the teacher, 'Could you please send Miss Cole down to the office.' I said, 'Oh my God, what have I done?' I was trying to remember what I did wrong... He said, 'Come in, Miss Cole... have a seat and relax and just have a seat and sit down.' I said okay. And so he says to me, 'I'm very proud of you.' I said, 'Why? What did I do?' He said, 'You're very bright and you're doing a very beautiful job... I'm going to put you on the honor roll... You keep up the good work. I know it's hard, but you're doing a wonderful job. You keep doing what you are doing.' So I remember that. That stuck with me."
101. Annie Cole (1956–1957), interview with the author, March 11, 2003.
102. Biographies accompanying senior pictures were examined in yearbooks for "tracking."
103. Judy Law (1950–1954), interview with the author, February 25, 2003: "Because I lived outside of town I had to depend on the bus to take me back and forth... it was very difficult for me to do anything after school and I usually didn't.... We had one car and my father worked seven days a week."
104. Annie Cole (1956–1957), interview with the author, March 11, 2003: "I worked after school. And I used to leave from school and walk to Emory Grove, and I used to clean the cottages over there... a couple of days a week... I was tired when I was finished because I had to walk back home... it's quite a walk. The lady used to give me eight dollars, and I used to give four dollars to my mom to put up and she let me have four dollars for my pocket. And at the end of the year, she'd take me downtown shopping." Doris Right (1956–1958), interview with the author, September 4, 2003: "We were very sheltered... we had to go home immediately after school... because my mother

had to work at night, and she wanted to make sure that we were home. And my father was a truck driver, so he wasn't home that much...She made sure we were in bed—dinner, homework, and bed."

105. This finding echoes educational ethnographer Michelle Fine's findings in the late 1980s of minority students dropping out proactively out of self-preservation, rather than out of failure. See Michelle Fine, *Framing Dropouts*.
106. Annie Cole (1956–1957), interview with the author, March 11, 2003: "Everybody would kind of stay away from the blacks...I couldn't, I just couldn't, I left....I just felt out of place at all times. I just didn't like it. My brother [one year older] liked it. My sister [four years younger] liked it...I graduated...from home."
107. Doris Right (1956–1958), interview with the author, September 4, 2003. Doris's memory of the details of the difference in curriculum between Miller and the all-black high school remained vague, while her emotional memory of the "different things" taught was strongly expressed in tone and inflection.
108. Doris Right (1956–1958), interview with the author, September 4, 2003: "All five of us [from the class of 1958] graduated."
109. Doris Right (1956–1958), interview with the author, September 4, 2003.
110. Norman Good (1956–1959), interview with the author, March 18, 2003.
111. Ibid.
112. Annie Cole (1956–1958), interview with the author, March 11, 2003: "My brother, he was the track star...he did a lot of things for the school. He was an honor roll student. He was just a bright kid. He really got along fine with everybody."
113. Norman Good (1956–1959), interview with the author, March 18, 2003.
114. Willie James and Jimmy Cole are deceased. Norman Good and Annie remembered them.
115. Football was considered too dangerous a sport in Baltimore County until the end of the 1960s, when it was first introduced.
116. Norman Good (1956–1959), interview with the author, March 18, 2003.
117. Ibid.
118. Norman Good (1956–1959), interview with the author, March 18, 2003: "[O]ne good friend, for the record, Frank Martell, we were on the track team together. He was white. We developed an excellent bond...he was the only person I let call me Norm. You know, we had that kind of a rapport and all. And he was a very popular individual...everybody knew Frank, because he had that kind of personality, he was a people person and all."
119. Norman Good (1956–1959), interview with the author, March 18, 2003.
120. Ibid.
121. The question remains why capable black young women did not make it into the academic track since they too would not have been recommended for a white higher education establishment after graduation. Did African American young women lack the parental support that would set them on a course toward higher education? Further research is required.
122. Norman Good (1956–1959), interview with the author, March 18, 2003.
123. Doris Right (1956–1958), interview with the author, September 4, 2003.

124. Ibid.
125. Nat Right (1958–1963), interview with the author, July 19, 2004.
126. Nat Right (1958–1963), interview with the author, July 19, 2004. Doris Right (1956–1958), interview with the author, September 4, 2003: "The one thing that really, really upset our family, probably the entire community...my brother had stayed at my grandmother's house who lived in Smithville. And the people that she worked for, the Griffiths, brought him to school...Because at that time, I don't think my grandmother had a telephone...the message didn't get to him. So he was dropped off at school, he walked around the corner, and he [a white male adolescent] hit him in the mouth with brass knuckles and he [her brother] lost all his front teeth...that was the worst thing that ever happened."
127. Annie Cole (1956–1957), interview with the author, March 11, 2003. The brother in question who hit the white boy was not Annie's popular brother. Annie also recalled: "The principal got all the blacks together, and he was very stern and said there wouldn't be any trouble. But we tried to explain to him...and we told him that the teachers saw it and didn't stop it. He said...some things he couldn't control."
128. Annie Cole (1956–1957), interview with the author, March 11, 2003.
129. Dotty Morris (1963–1965), interview with the author, September 24, 2004.
130. Dotty Morris (1963–1965), interview with the author, September 24, 2003.
131. Ibid.
132. Dotty Morris (1963–1965), interview with the author, September 24, 2003: "I was involved and I liked to sing and so did my girlfriend, so we were always singing. We were in the glee club and the choir when I was in middle school. I was in music also there. [At Miller High] I was in the school orchestra. I played the violin."
133. Dotty Morris (1963–1965), interview with the author, September 24, 2003.
134. Dotty Morris (1963–1965), interview with the author, September 24, 2003: "Mary and Thomas were in my class."
    "Mary was white and Thomas was black?"
    "Yeah. And they're still married today, three kids."
    "Wasn't it unusual to see an interracial couple then?"
    "It was kind of the start, you know, at that time."
    "Were they shunned by people in the community?"
    "I don't really know about that, but I think that there were some, you know what I mean about that. But they [Mary and Thomas] continued on, and I don't know what it was that she heard, because, you know, I didn't go there."
135. Dotty Morris (1963–65), interview with the author, September 24, 2003.
136. Ibid.
137. Dotty Morris (1963–65), interview with the author, September 24, 2003.
138. Ibid.
139. Annie Cole (1956–1957), interview with the author, March 11, 2003: "My sister liked it [Miller High]...She's younger than me—four years."
    "Did she graduate or just attend?"
    "She graduated.... She's a nurse."

140. Nat Right (1958–1963), interview with the author, July 19, 2004: "Jocks were popular."
    "Were you a jock?"
    "No. When I came through, there was James and Samuel. They were good sportsmen...They gravitated to the jocks, the whites and the black jocks."
141. Nat Right (1958–1963), interview with the author, July 19, 2004.
142. Ibid.
143. Nat Right (1958–1963), interview with the author, July 19, 2004.
144. Burt Sadden (1963–1967), interview with the author, July 26, 2004.
145. Ibid.
146. Burt Sadden (1963–1967), interview with the author, July 26, 2004: *"Did you participate in extracurricular activities?"*
    "No...it was go to school and work. After school it was always try to find part-time work, to help the family, and to survive...a few would stay, in the higher group, or the students, I should say, in the more intelligent classes seemed to hang around and do more at the school than just the general students. Most of the students did their hours and then got on the bus."
147. Burt Sadden (1963–1967), interview with the author, July 26, 2004.
148. Burt Sadden (1963–1967), interview with the author, July 26, 2004: "Blacks stayed to themselves and whites stayed to themselves."
    "Was that because of racial animosity?"
    "Pretty much culturally."
149. Burt Sadden (1963–1967), interview with the author, July 26, 2004.
150. Nat Right (1958–1963), interview with the author, July 19, 2004.

2 Cautiously Negotiating Social Divides: A Conservative Student Body (1950–1969)

1. Judy Law (1950–1954), interview with the author, February 25, 2003: "We were asked out by boys who were at another school...And I got asked out by a couple of people who were, I would say, out of my league...meaning their family had a lot of money...And I remember two in particular who asked my father to take me out and he said no...He just didn't like the idea. I remember telling my father later...you know, you may have just kept us from having the family fortune."
2. Alice Web (1950–1954), interview with the author, October 2, 2003.
3. Bud Land (1965–1969), interview with the author, August 16, 2003.
4. Alice Web (1950–1954), interview with the author, October 2, 2003.
5. Robert Heart (1952–1956), interview with the author, June 24, 2003: "We couldn't date the upper-class girls. They just wouldn't date us. They wouldn't. You would ask them for a date and they would just turn around and walk away. [But] if you were a junior [upper-class woman], you could date a senior of the lower class."

6. Robert Heart (1952–1956), interview with the author, June 24, 2003.
7. Norman Good (1956–1959), interview with the author, March 18, 2003.
8. Nat Right (1956–1963), interview with the author, July 19, 2004.
9. Doris Right (1956–1958), interview with the author, September 4, 2003: "My parents had bought land on the Corner of C. Avenue...." The Rights had moved to a wealthier white neighborhood. As discussed in the Introduction, black families of Miller Town of the 1950s and 1960s owned their own businesses; moreover, these families' roots reached back into the nineteenth century and their mobility within Miller Town would prove relatively easy given their long-established standing within the majority white community.
10. White and black, female and male graduates reported attending either the local Episcopal, Catholic, or Methodist churches.
11. Annie Cole (1956–1957), interview with the author, March 11, 2003: "They [white students] were throwing Ivory Soap on the floor...I thought they were maybe trying to make the floor slippery so we could dance better. You know, get more sliding and doing. But no, ma'am...it was to instigate a fight. I didn't understand what was really going on...I thought we [the black students] just want to dance...So it finally escalated. Somebody hit somebody. Somebody called somebody that bad name."
12. See Table 3 in Addendum, Methodology: The Transparent Historian. See also Thomas J. Vicino, *Transforming Race and Class in Suburbia: Decline in Metropolitan Baltimore*, 42.
13. Annie Cole (1956–1957), interview with the author, March 11, 2003: "They ran us out of the school." Teachers' tacit support of white students' rejection of blacks on the dance floor was most likely also a reflection of their fear of black expression. Historian Grace Palladino explains that "Rock 'n' roll was everything that middle-class parents feared: elemental, savage, dripping with sexuality, qualities that respectable society usually associated with depraved classes." See Grace Palladino, *Teenagers: An American History*, 155.
14. Linda Eisenman, *Higher Education for Women in Postwar America, 1945–1965: Reclaiming the Incidental Student* (Baltimore: Johns Hopkins University Press, 2006), 19.
15. Ibid, 18.
16. See Paula Fass, *The Damned and the Beautiful: American Youth in the 1920s* (NY: Oxford University Press, 1977); Barbara Finkelstein, "Is Adolescence Here to Stay? Historical Perspectives on Youth and Education," in *Adolescence and Society*, eds., T. Urban and F. Pajares (NY: Information Age Press, 2003), 1–33.
17. Eisenman, *Higher Education for Women in Postwar America, 1945–1965: Reclaiming the Incidental Student*, 18.
18. It is worth noting that none of the white female narrators reported aggressive behavior by males toward them or toward other white women they may have recalled. All alumni were asked to describe their perceptions of and relationships with the opposite sex at high school. More research is needed to see if white adolescent girls were in fact less likely than their black counterparts to be sexually harassed by white adolescent boys.

19. Doris Right (1956–1958), interview with the author, September 4, 2003.
20. Franklin Frazier, *Negro Youth at the Crossroads: Their Personality Development in Middle States* (NY: Scholar Books, 1967), 105. See also Barbara Finkelstein, cited earlier.
21. Annie Cole (1956–1957), interview with the author, March 11, 2003.
22. Annie Cole (1956–1957), interview with the author, March 11, 2003: "I played girls' softball, which was fun. You know, when you're playing sports, everybody [narrator smiled remembering good times], yeah, now that was fun." Nora Jones (1955–1959), interview with the author, June 12, 2003: "Rose [African American peer], who played hockey, you just didn't want to meet that girl coming down the hockey field [narrator smiled with appreciation for Rose's skills]."
23. Annie Cole (1956–1957) excelled in gym; Robert Heart loved drawing in shop classes, and after years of diverse occupations (from soldier in the army to working for an architectural firm), became an architect without ever having attended college; Nora Jones became an artist, and raved about all the art teachers, etc.
24. See works by Samule Bowles and Herbert Gintis, *Schooling in Capitalist America* (NY: Basic Books, 1976). Yet, against socially reproductive norms, testimonies often revealed the importance of relationships with teachers who cared for youth across race and class divides; relationships with teachers that changed or helped the lives of alumni—including African American Norman Good, who went on to hold prestigious positions in corporations and the federal government. He recounted having been able to land better jobs in mostly white settings because of the vote of confidence he had received from his math and foreign language teacher at Miller High. He felt confident succeeding among white people.
25. Sherry Parson (1958–1962), interview with the author, March 7, 2003.
26. See Finkelstein, "Is Adolescence Here to Stay? Historical Perspectives on Youth and Education"; and Fass, *The Damned and the Beautiful: American Youth in the 1920s*.
27. Almost all graduates across gender and race categories reported a social stigma attached to "dropping out" and parents' full expectation that their children would graduate from high school. All graduates—white and black, female and male—commented on the importance their parents placed on their high school education. For white and black males in the general track in particular, who were "doing time," it was their family members' insistence on a high school education and their own loyalties to their family that kept them in school.
28. At the dawn of the twentieth century, G. Stanley Hall, president of Clark University, and Charles W. Elliot, president of Harvard University, debated about the direction American high schools should take. The view that all students' needs should be addressed through a diversified curriculum became the template for the comprehensive high school. See G. Stanley Hall, "How Far Is the Present High School and Early College Training Adapted to the Nature and Needs of Adolescents?" *School Review* 9 (1901):

649–681. The report *Cardinal Principles of Education*, published by the Bureau of Education, U.S. Department of the Interior (Washington, D.C.: Government Printing Office, 1918), fully espoused Hall's argument on behalf of a diversified curriculum to keep as many students involved in high school life as possible.
29. James Coleman, *The Adolescent Society: The Social Life of the Teenager and Its Impact on Education* (NY: Free Press of Glencoe, 1961).
30. James Coleman, *The Adolescent Society: The Social Life of the Teenager and Its Impact on Education*, 217.
31. See Finkelstein, "Is Adolescence Here to Stay: Historical Perspectives on Youth and Education;" see also Robert Lynd and Helen Lynd, *Middletown* (NY: Harcourt Brace, 1929).
32. Status titles such as homecoming queen and May queen, as well as many of the extracurricular activities, were not student-generated in the 1950s and 1960s, but were handed down to students as integral parts of the school's modus operandi, further underscoring the weight of school structure in the lives of the Divided Generation. These were institutionalized performances.
33. Linda Moss (1965–1969), interview with the author, September 24, 2003.

3 Memories of Interracial Peer-Group Affiliations: Integration Years(1970–1985)

1. Joanne Pet (1970–1974), interview with the author, October 16, 2003.
2. Tim Whittle (1977–1981), interview with the author, June 28, 2004.
3. Tim Whittle (1977–1981), interview with the author, June 28, 2004. Also, Jeremy Garnes (1972–1976), interview with the author, September 9, 2003: "African Americans at the time were all local. They were local Miller Town people that we knew from growing up with." The term "local" is often used by Miller Town residents who have long roots in the community during casual conversations among neighbors.
4. See Table 3 in Methodology: The Transparent Historian.
5. Michelle Shaw (who began teaching at Miller High in 1960), interview with the author, November 7, 2003.
6. African American Teresa Randle (1977–1981), interview with the author, June 9, 2004.
7. Nat Right is the one graduate of the Divided Generation who referred to athletes of his time as jocks. The jock and the cheerleader can be identified as early as the 1920s, when football became an institutional feature of secondary education. See works by Paula Fass, *The Damned and the Beautiful: American Youth in the 1920s* (NY: Oxford University Press, 1977), and Elliott West, *Growing Up in Twentieth-Century America: A History and Reference Guide* (Westport, CT: Greenwood Press, 1996). The absence of football at Miller High until end of the 1960s is a cultural anomaly.

8. Jeremy Garnes (1974–1978), interview with the author, September 9, 2003: "I was a jock. That was the big thing. That's what most of us cared about." Also, David Randle (1972–1976), interview with the author, November 7, 2003: "We were jocks...it's [playing football] something I really wanted to do."
9. Tim Whittle (1977–1981), interview with the author, June 6, 2004. Also, David Randle (1972–1976), November 7, 2003: "I knew everybody!" Also, Jeremy Garnes, (1974–1978), September 9, 2003: "My first day was actually not at school. It was on the field."
10. Josh White (1972–1976), interview with the author, October 1, 2003.
11. Tim Whittle (1977–1981), June 6, 2004.
12. Ibid.
13. David Randle (1972–1976), November 7, 2003.
14. Ibid.
15. Ibid.
16. Ibid.
17. Ibid.
18. Most graduates interviewed, whether black or white, male or female, identified the fully integrated experience of the black jock at Miller High. The musician Michael Hallner (1972–1976), interview with the author, September 2003, explained: "If the black students were jocks, they stayed with [white] jocks."
19. David Randle (1972–1976), November 7, 2003.
20. Ibid.
21. Ibid.
22. Ibid.
23. Tim Whittle (1977–1981), interview with the author, June 28, 2004. Also, Jeremy Garnes (1974–1978), interview with the author, September 9, 2003: "My education was probably not the best at Franklin High. I knew it was too easy...today [my kids'] workload is very, very hard....I don't believe I had any homework that I remember...When I went to college it was a rude awakening."
24. Tim Whittle (1977–1981), interview with the author, June 28, 2004: "The rules were breaking down...We had release time, which means if you had enough credits, you didn't have to take certain amount of classes. I didn't take seventh period at all senior year. And one whole school day, I just had study period."
25. Historians have tracked this trend among white male students since the early decades of the twentieth century. See Paula Fass, *The Damned and the Beautiful: American Youth in the 1920s*.
26. Gerald Grant, in *The World We Created at Hamilton High*, alludes to teachers' laissez-faire practices during the 1970s at Hamilton High, a city high school. However, he attributes those lax practices to younger teachers' "guilty liberalism," teachers' general confusion about standards and discipline regarding black students under pressures of advocacy groups, and redefinitions of relationships between minors and adults with Supreme Court decisions such as the *Gault* case in 1967 and the *Winship* case in 1970. Of interest here is that

academic leniency was practiced at Miller High not by younger, but by older teachers who could not wait to retire.
27. Jeremy Garnes (1974–1978), September 9, 2003. Echoing Jim Garnes'experience, Tim Whittle (1977–1981) recalled: "It was two sets of teachers. You got your young ones and your older ones... You could go right down my grades and see who my teacher was... the older teachers were putting in their time."
"The older ones were the easy grade?"
"The older ones because they didn't give a hoot anymore. The younger ones challenged you."
28. David F. Labaree, *How to Succeed in School without Really Learning* (New Haven: Yale University Press, 1997).
29. Josh White (1972–1976), interview with the author, October 1, 2003.
30. Jeremy Garnes (1974–1978), interview with the author, September 9, 2003.
31. David Randle (1972–1976), interview with the author, November 11, 2003: "He [the coach] saw something in me and he helped me out. He really helped me a lot." Also, Tim Whittle (1977–1981), interview with the author, June 6, 2004: "He [the coach] was one of the nicest, best teachers I ever had. He and I really got along... my graduation, he made it a point to find my father and told my dad that I was one of the best students he had. That meant a lot to me... He was always in my corner." Also, Josh White (1972–1976), interview with the author, October 1, 2003: "Everybody liked him [the coach]."
32. Josh White (1972–1976), interview with the author, October 1, 2003.
33. African American alumna Teresa Randle (1977–1981), interview with the author, June 9, 2004.
34. Teresa Randle (1977–1981), interview with the author, June 9, 2004.
35. Sophie Baker (1981–1985), interview with the author, June 17, 2004.
36. Teresa Randle (1977–1981), interview with the author, June 9, 2004.
37. Sophie Baker (1981–1985), interview with the author, June 17, 2004.
38. Teresa Randle (1977–1981), interview with the author, June 9, 2004:
"I keep in touch with my teammates to this day!"
"A lot of bonding happens in sports?"
"It does!"
39. Of all the graduates interviewed during this period, Sophie, a 1985 graduate, was the only one to speak about the tangible markers of her life as a high school jock. Needless to say, only many more interviews might reveal whether this is a gender-specific recollection more akin to white female students, or if students of the 1970s were generally less interested in material possessions than Sophie, who graduated in the mid-1980s, might have been, reflecting a more materially conscious time. Then, too, Sophie's recollection might simply reflect her particular personality. Nevertheless, such details capture the known trend among U.S. high school students since the 1930s/1940s bobbysoxers of mass consumption of high-school-specific goods, from class rings to pens to pins. See Grace Palladino, *Teenagers: An American History*.
40. Teresa Randle (1977–1981), interview with the author, June 9, 2004: "If I had studied, I could have done better. I was just content with getting

by... Bs and Cs." Also, Sophie Baker (1981–1985), interview with the author, June 17, 2004: "I would actually calculate how much homework I could miss and still get a B... there were times when I didn't feel like doing it. I played sports."
41. Teresa Randle (1977–1981), interview with the author, June 9, 2004: "I didn't take advantage of the opportunity that was there. I did enough to get by." Also, Sophie Baker (1981–1985), interview with the author, June 17, 2004: "In retrospect... it was easy. Because it was easy, I didn't do everything I should have done."
42. Teresa Randle (1977–1981), interview with the author, June 9, 2004: "They didn't tell us about other colleges."
43. Sophie Baker (1981–1985), interview with the author, June 17, 2004.
44. Josh White (1972–1976), interview with the author, October 1, 2003: "When I was in school, you were either a head or a jock."
45. Ibid.
46. Jeremy Garnes (1974–1978), interview with the author, September 9, 2003.
47. Josh White (1972–1976), interview with the author, October 1, 2003.
48. Ibid. Josh White's use of the term "funny" was not expressed in a derogatory manner as if to laugh at the mailman, but in a manner that suggested surprise over the fact that the mailman is a peer who attended high school when he did. Nevertheless, the particular recollection suggested a "wasted" life.
49. Tim Whittle (1977–1981), interview with the author, June 28, 2004: "a buddy of mine, a football player, and he loved to get high. The teacher said, 'I'm going to make sure you don't graduate, you're a loser.'"
50. Sam Garnes (1970–1974), September 16, 2003.
51. A stereotype reinforced in popular movies such as *Grease*, where the car is used for "making out," engaging in illegal drag racing, or getting high.
52. From an informal conversation with Sam Garnes (1970–1974). Sam shared this information during our informal conversation that followed the tape-recorded conversation. Within about thirty-five minutes of recording, Sam asked that we turn off the recorder and continue our conversation without being recorded.
53. Joanne Pet (1970–1974), interview with the author, October 16, 2003
54. Carnoy and Levin quoted in David F. Labaree, *How to Succeed in School without Really Learning* (New Haven: Yale University Press, 1997), 49.
55. Pat Baley (1971–1975), interview with the author, October 20, 2003: "I was a musician... I had my own band, classic rock." Also, Michael Hallner (1972–1976), interview with the author, September 2003: "We [Michael and two other friends] got together, practiced, but never got it rollin' 'cause we all had a difference of opinion of music. They wanted to play Kiss... I wanted to play top forty, where you can make some money."
56. Tim Whittle (1977–1981), interview with the author, June 28, 2004.
57. Linda Moss (1965–1969), interview with the author, September 9, 2003.

4 Bridging Social Divides Through Peer-Groups: A Socially Tolerant but Politically Inactive Student Body (1970–1985)

1. Speaking of cross-racial relationships: African American alumnus Pat Baley ((1971–1975), interview with the author, October 20, 2003) said: "We got along. Most of the students you went to high school with, you started with in elementary school." Also, white alumnus Josh White, (1972–1976), interview with the author, October 1, 2003: "We got along really well." African American alumnus David Randle, (1972–1976), interview with the author, November 7, 2003: "My best friends were white." African American alumnae Joanne Pet and Teresa Randle and white alumna Sophie Baker reported as much.
2. Refer to African American alumna Teresa Randle's testimony in chapter 3 regarding her white friend and her KKK father, as well as to stories by African American alumnus David Randle and white alumnus Tim Whittle regarding white peers policing other white peers against racist attitudes.
3. See the work of Philip Cusick, *Inside High School: The Student's World* (NY: Holt, Rinehart, and Winston, 1973). Cusick argues that the small friendship groups of youth are reactions to the disempowering effects of secondary schools, which herd youth and devoid them of autonomy.
4. See Stephanie Coontz, *The Way We Never Were: American Families and the Nostalgia Trap* (NY: Basic Books, 2000), xi.
5. Tim Whittle (1977–1981), interview with the author, June 28, 2004.
6. Speaking of their principal, white alumnus Josh White (1972–1976), interview with the author, October 1, 2003: "I don't remember him when I was there." Also, African American alumna Teresa Randle (1977–1981), interview with the author, June 9, 2004: "Mr. Krauss was distant. I mean, he didn't really know me personally. He was distant."
7. The football jock culture might have participated in shifting aggression from male students against coaches (the reader will recall Robert Heart's testimony of his friend punching the coach and dropping out) to aggression by coaches against male students. For accounts of connections between sports and racial integration, see Pamela Grundy's acclaimed work, *Learning to Win: Sports, Education, and Social Change in Twentieth-Century North Carolina* (Chapel Hill: University of North Carolina, 2001).
8. A series of laws had been passed that expanded students' rights in secondary schools as well as the rights of young people in juvenile court proceedings. See rulings in *Tinker vs. DeMoines*, 1968; also in 1968, the ACLU published "Academic Freedom in the Secondary Schools," in which ACLU lawyers argue for "a recognition that deviation from the opinions and standards deemed desirable by the faculty is not ipso facto a danger to the educational

process"; see also Supreme Court decisions in the *Gault* case in 1967 and the *Winship* case in 1970. For details about *Tinker vs. DeMoines*, see Urban and Wagoner,*American Education: A History* (NY: McGraw Hill, 2000), 338. For details about the ACLU on academic freedoms in secondary education, and the *Gault* and *Winship* cases, see Gerald Grant, *Hamilton High*, 50–51.

9. Speaking about younger teachers: Tim Whittle (1977–1981) (interview with the author, June 28, 2004): "The younger ones challenged you. They were the ones you got into arguments with. They also stimulated you... If they had to put up with you being a bit playful or something, as long as it did not get out of hand, sometimes they would join in, they would find it funny."
10. Teresa Randle (1977–1981) (interview with the author, June 9, 2004): "I always had a good rapport with teachers." Also, Sophie Baker (1981–1985) (interview with the author, June 17, 2004): "I think a lot of people formed fairly close bonds with teachers at that time."
11. Sophie Baker (1981–1985), interview with the author, June 17, 2004.
12. Refer in part I, chapter 1, to Budd Land's remark (1964–68): "The general track was for those who were just going to go out and work as laborers. They were the slower kids or those that just didn't care about what they were doing."
13. Tim Whittle (1977–1981) reported other incidents of teachers calling students losers (see chapter 3). Pothead Sam Garnes (1974) and musician Michael Hallner (1986) also mentioned teachers' divisions of students into winners and losers.
14. For a succinct analysis of the social-efficiency perspective on education, see Chapter 1 in Labaree, *How to Succeed in School without Really Learning* (New Haven, CT: Yale University Press, 1997), 15–52.
15. Refer to Philip Cusick, *Inside High School: The Student's World*, earlier cited.
16. See Arthur G. Powell and Eleanor Farrar, *The Shopping Mall High School* (Boston: Houghton Mifflin, 1985), as well as Labaree's work earlier cited.
17. Kenneth Fish, *Conflict and Dissent in the High School*, 6.
18. E.F. Frazier, *Negro Youth at the Crossroads: Their Personality Development in the Middle States*, 105.
19. For works that examine black high school youth political engagement, see Dwayne C. Wright, "Black Pride Day, 1968: High School Student Activism in York, Pennsylvania," *Journal of African American History* 88 (2003): 151–162; Dionne Danns, "Chicago High School Students' Movement for Quality Education, 1966–1971," *Journal of African American History* 88 (2003): 138–150. See also V.P. Franklin, "Black High School Student Activism in the 1960s: An Urban Phenomenon?" *Journal of Research in Education* 10 (Fall 2000): 3–8.
20. Joanne Pet (1970–1974), interview with the author, October 16, 2003.

5 Memories of Segregation by Class, Race, Nationality, and Religion: Destabilizing Years of Shifting Demographics (1986–2000)

1. Heather Korn (1989–1993), interview with the author, February 25, 2003.
2. Sue Cohen (1991–1995), interview with the author, September 2003.
3. <Miller High> Alumni Directory (2000), xi.
4. At that time, many Asian Indians and Mexicans were also immigrating to Miller Town, but in much smaller numbers than Russian Jews (see Table 3). Because of limited resources available to conduct this oral history research and the larger number of Russian immigrants, I chose to concentrate my efforts on locating Russian alumni. See Addendum: Methodology.
5. <Miller Town> Community Times, throughout the 1970s. Holdings in archives of <Miller Town> library.
6. <Miller Town> Community Times, throughout the 1970s. Holdings in archives of <Miller Town> Library. Also, alumni recalled the transformations they had witnessed: Harry Rice (1983–1987), interview with the author, February 25, 2003: "It was strange to see so many new houses and roads that weren't there before." Cherry Gate (1988–1992), interview with the author, October 23, 2003: "My father, he's seen a complete change in the community since he's been there, and he's been there for thirty-one years. Yeah. And they built up around us...I mean, it used to be like all farmland...I forget about the malls...'cause Springfield Mall was just opening and Oak Field at the time."
7. Vera Debin (1995–1999), interview with the author, January 15, 2004.
8. Also, African American alumna Teresa Randle, 1981 graduate, compared the 1990s black student population to that of the 1970s: "It has changed now. I went to a graduation last week, there are more African American kids, almost 60 percent. I read in the paper that they have more racial tensions. I didn't witness that when I was in school [in the 1970s]."
9. David Randle (1972–1976), interview with the author, November 7, 2003. These reflections on changes in the community were shared by those alumni who were themselves "old-timers," and most had graduated from Miller High in the 1970s and early 1980s. David Randle, who is still a resident of Miller Town, has witnessed changes in the community firsthand.
10. Tim Whittle (1977–1981), interview with the author, June 28, 2004. Again, Tim Whittle counts among the "old-timers." There is a parallel here to the reaction of northern city blacks who felt that their good standing among the white community was endangered by the arrival of southern rural blacks. See Nicholas Lemann, *The Promised Land: The Great Black Migration and How It Changed America* (NY: Alfred A. Knopf, 1991).
11. African American alumna Teresa Randle, 1981 graduate, explained: "I have one child, she goes to a private school. So long as I can do it, I will keep her

there. I feel it is safer...There are a lot of kids that are angrier than ever before. Other people are raising them...We have kids in group homes in this area now...It's like foster homes, which means that more and more kids are on medication. When they don't take their medication you don't know what kind of child you will have that day. I just don't want my daughter to be involved in that." African American alumnus Pat Baley explained: "We homeschool. My kids are good and I want to keep them that way. Kids get all their bad habits starting in middle school. [I want] to protect [my kids] against bad influence of peers." White alumna Roberta Jones, 1991 graduate, explained her parents' choice regarding the schooling of her brother: "My brother went to a private school...[My parents] were worried the years that came under me, each year seemed to be a little bit worse and worse, worrying about whom your kids were hanging out with...He was a boy and he could get into more trouble."

12. Harry Rice (1983–1987), interview with the author, February 25, 2003. While the traditional ringing of the bell on the first day of school continued, and while the building remained, in the eyes of the Redivided Generation, an antediluvian construction, students' means and styles of communication were high tech. Administrators incorporated into school policies rules about the use of things such as cell phones. In the decade of Columbine-like tragedies, because of the insistence by many parents that their children be allowed to use cell phones on school premises, both as emergency tools as well as means for coordinating last-minute changes in overbooked schedules, and because of the speed at which drug deals could be arranged between students during school hours, and friends or lovers could chat away instruction time across classrooms through text messages, the cell phone emerged as "a tool so great," to quote the humorous Firesign Theater, "it could only be used for good or evil." Conference calls and text messaging further extended young people's ability to connect with each other in virtual communities. In the three-dimensional, physical world, Miller High School students now met at the recently opened mall in a neighboring town.

13. Heather Korn (1989–1993), interview with the author, February 25, 2003: "[There was] definitely kissing in the hallways...When you've seen it that often—we'd seen it since seventh and eight grade—they started way back. A girl, the summer of eighth grade, had twin boys...she was twelve actually. So we have been seeing it for a while."

14. Heather Korn (1989–1993), interview with the author, February 25, 2003: "If there was an alcohol or cigarette advertisement, they would be told to turn your shirt inside out, or they would have somebody come get you or bring you new clothes or something like that."

15. Ibid: "There was one kid, one or two classes ahead of us, who was very openly gay, and he got teased a lot...he got it a lot."

16. A member of FCA and of Young Life Club during his high school career, white alumnus Bill Jackman (1995–1999) (interview with the author, October 10, 2003) said, "We have a bible study." Also, Russian immigrant Vera Debin (1995–1999) (interview with the author, January 15, 2004): "Ultra right-wing [students attended] the morning bible study."

17. Of interest is that while gender no longer featured as a primary organizing category for 1990s alumni who consistently reported equal treatment of, interaction between, and participation by females and males, sexuality emerged as an organizing category. However, it was only briefly alluded to by one alumna.
18. All alumni interviewed had been "rule-abiding" students. While I was able to identify a student who had counted among those I have labeled "rule breakers," the alumnus refused to be interviewed or to suggest other alumni for me to interview. Thus "rule breakers" in this study come into view indirectly through recollections of those alumni interviewed, as well as through the recollections of two teachers whose oral histories I also collected and whose tenures reached back to the early 1960s through the early 1970s. See Table 1.
19. The following quotes illustrate the progressive decline of school spirit as we reach the end of the century. Harry Rice (1983–1987) (interview with the author, February 25, 2003): "[P]arents were involved...there was a lot of spirit." Also, Cherry Gate (1988–1992) (interview with the author, October 23, 2003): "We showed our school spirit...the band, and the choir, the auditorium was full. There was a lot of family, alumni who used to come and watch." Cecilia Hood (1992–1996) (interview with the author, October 7, 2003): "[School spirit] during homecoming week, but other than that, not really." Bill Jackman (1995–1999) (interview with the author, October 18, 2003): "I wouldn't say it was...school-spirited."
20. Sue Cohen (1991–1995), interview with the author, September 2003.
21. Bill Jackman (1995–1999), interview with the author, October 18, 2003.
22. Sue Cohen (1991–1995), interview with the author, September 2003.
23. As the Jewish student population grew at Miller High in the 1990s, more Jewish names appeared in the yearbook under pictures of the journalism club, suggesting a continuation into the end of the century, as the historian Paula Fass had identified, of proportionately more Jews, in particular female Jews, in extracurricular activities involving writing. See Paula in Joe Austin and Michael Willard, earlier cited.
24. Bill Jackman (1995–1999), interview with the author, October 18, 2003.
25. Bill Jackman (1995–1999) (interview with the author, October 18, 2003): "I remember one of my honors history teachers that was very demanding. We had to write a report on two presidents a week—four, five pages each president."
26. Sue's testimony underscores once again the weight of economic status in academic competition, given that attending college summer courses requires financial wherewithal.
27. Scans of yearbook captions next to senior pictures across time periods consistently reveal proportionately fewer black students in honors and gifted classes, which were referred to by earlier generations as "academic" tracks.
28. Although I was not able to secure interviews with several black students who graduated in the 1990s (see Addendum: Methodology) and whom I had identified through snowballing, I know firsthand through my personal friendship with many long-time black residents of Miller Town that many of the young black people in the community were and continue to

be actively involved in their churches. Russian immigrant alumna Vera Debin, 2000 graduate (interview with the author, January 15, 2004), recalled "black kids speaking of Jesus" in the hallways and talking about church meetings.
29. Ivan Strasky (1998–2002), interview with the author, November 6, 2004.
30. The guidance counselor at Miller High spoke highly of Ivan and recommended that I interview him to capture the Russian perspective.
31. Vera Debin (1995–1999) (interview with the author, January 15, 2004): "There was a very few teachers that I had any kind of relationship with other than in class." Cecilia Hood (1992–1996) (interview with the author, October 7, 2003): "I don't really remember the teachers that much."
32. Harry Rice (1983–1987) (interview with the author, February 25, 2003): "I was one of the geeks, no doubt."
33. Bill Jackman (1995–1999) (interview with the author, October 18, 2003): "The Russian kids were definitely a group." Cecilia Hood (1992–1996) (interview with the author, October 7, 2003): "People kind of stayed in groups...we had a large population of Russian kids." Betty Ames (1995–1999) (interview with the author, August 19, 2003): "Everyone definitely segregated themselves...two tables of...and then the Russian kids."
34. Ivan Strasky (1998–2002) (interview with the author, November 6, 2004): "Some [Russian] kids were athletes...I do remember two Russian kids who did wrestling."
35. Vera Debin (1995–1999), interview with author, January 15, 2004.
36. Ivan Strasky (1998–2002) (interview with author, November 6, 2004): "When Russian kids chose not to take the path of defensiveness, they [could] easily succeed." Vera Debin (1995–1999) (interview with author, January 15, 2004): "The programs were not difficult to get into; anyone who put in the effort would be able to do so. The people that were in standard, they chose to be there. They didn't want to put the extra effort. They were satisfied with their place."
37. Ivan Strasky (1998–2002), interview with the author, November 6, 2003.
38. Vera Debin (1995–1999), interview with the author, January 15, 2004.
39. Cherry Gate (1988–1992), interview with the author, October 23, 2003.
40. Heather Korn (1989–1993), interview with the author, February 25, 2003. Mr. L. presided over Miller High between 1978 and 1991, and his female successor, the only female principal in the history of Miller High during the twentieth century, presided between 1991 and 1997.
41. Bill Jackman (1995–1999), interview with the author, October 18, 2003.
42. Cherry Gate (1988–1992), interview with the author, October 23, 2003.
43. Bill Jackman (1995–1999) (interview with the author, October 18, 2003): "There were people in school that were racist either way."
44. Cecilia Hood (1992–1996), interview with the author, October 7, 2003.
45. Cherry Gate (1988–1992), interview with the author, October 23, 2003.
46. Heather Korn (1989–1993), interview with the author, February 25, 2003.
47. Heather Korn (1989–1993) (interview with the author, February 25, 2003): "All my friends were in standard."

48. Heather Korn (1989–1993) (interview with the author, February 25, 2003) echoed Cherry Gate's assessment of preps: "A lot of the kids that did hang out with the kids that did have the money, they would have the snobby attitude."
49. Heather Korn (1989–1993), interview with the author, February 25, 2003. Although Heather did not directly identify the group, and although it is impossible to extract a quote from the transcript which would confirm it, there is good reason to believe that she spoke of the Russian students, since she had identified all other groups by their names directly (blacks, preps, etc.), but had not mentioned the Russians. All other graduates of the Redivided Generation directly identified the Russians.
50. Cherry Gate (1988–1992) (interview with the author, October 23, 2003): "I worked for Riverside Veterinary Hospital my senior year...well, I used to groom when I was at horse shows, and I worked at my parents' farm." Roberta Jones (1987–1991) (interview with the author, September 2, 2003): "I worked as a grocery clerk on weekends and a couple of times during the week."
51. Heather Korn (1989–1993) (interview with the author February 25, 2003): "I cried the entire time...I'm pretty sentimental...I had a friend in every row...some of them knew they were going in the service...I had one friend whose dream it was to be a housewife." Cherry Gate (1988–1992) (interview with the author October 23, 2003): "Graduate. Move on with my life."
52. Roberta Jones (1987–1991), interview with the author, September 2, 2003.
53. Cherry Gate (1988–1992) (interview with the author, October 23, 2003): "A lot of the guys I hung out with were EMT and junior firefighters." Cherry Gate also recalled that her then-boyfriend, who is now her spouse, had been in the work-release program, and that he trained in a local mechanic shop. Heather Korn (1989–1993) (interview with the author, February 25, 2003): "The junior firefighters. You heard the siren go off, they all sat up straight, itching to get out of their seats."
54. Cecilia Hood (1992–1996), interview with the author, October 7, 2003.
55. Betty Ames (1995–1999), interview with the author, August 19, 2003. Also, Cecilia Hood (1992–1996) (interview with the author, October 7, 2003): "The popular kids, some of us played sports...we were friends since middle school."
56. Also Vera Debin (1995–1999) (interview with the author, January 15, 2004): "It was big limousines and big fancy dresses."
57. See Finkelstein earlier cited.
58. Cecilia Hood (1992–1996), interview with the author, October 7, 2003.
59. Sue Cohen (1991–1995), interview with the author, September 2003.
60. Vera Debin (1995–1999, interview with the author, January 15, 2004.
61. Betty Ames (1995–1999) (interview with the author, August 19, 2003): "I would sit in front of the TV from eight to eleven every night doing my homework. I'd do it in front of the TV. I didn't make straight As, I got by...When I went to college, I went *wow* [speaker's emphasis]...freshman year I did more work than I ever did in all four years of high school." Also Cecilia Hood

(1992–1996) (interview with the author, October 7, 2003): "I don't remember the teachers that much. I remember the social and the sports."
62. Refer to Addendum, Methodology: The Transparent Historian.
63. Vera Debin (1995–1999), interview with the author, January 15, 2004. I asked Vera, "Would black kids attend those [bible study] meetings?" Vera categorically answered, "No." I asked, "So this was just—," at which point Vera interrupted to say, "The ultra-religious right."
64. Sue Cohen (1991–1995) (interview with the author, September 2003): "Our football team wasn't so good." Also, Cecilia Hood (1992–1996) (interview with the author, October 7, 2003): "Football was horrible back then." Also Bill Jackman (1995–1999), interview with the author, October 18, 2003: "The football team didn't get a whole lot of support because they were one and nine every year while I was in high school."
65. Betty Ames (1995–1999) (interview with the author, August 19, 2003): "The soccer players were white, but there'd be an Indian kid who'd play soccer." This was corroborated by yearbook data. Further research should investigate the reasons for lack of black male students' participation in lacrosse and soccer.
66. For a discussion of the manifestations of punk subculture, from suburban youth whose encounters with punk are "primarily through videos on MTV and VH1" to the "genuine punks...who listen solely to independently produced anarcho-punk, steal or panhandle their means of subsistence...and eschew most forms of commodity," see Stacy Thompson, "Punk Cinema," *Cinema Journal* 43 (2004): 47.
67. Ibid.
68. For a discussion of the German roots of the goth subculture that emerged in the 1990s, see Gabriele Eckart, "The German Gothic Subculture," *German Studies Review* 28 (2005): 547–562. In this work, Eckart discusses a continuum among goths, who, while generally nonpolitical, have exhibited tendencies toward fascism at one extreme and antimilitarism at the other.
69. For a brief review of Afro-punk, see Black Grooves, "Afro-Punk," Archives of American Music and Culture, http://blackgrooves.org/?p=283 (accessed July 17, 2009).
70. Ivan Strasky (1998–2002), interview with the author, November 11, 2004.
71. David Randle (1976) alluded to African American pride when he stated that the new generation was no longer interested in cooperating, but rather in seeing faster change in the white status quo. Russian pride was alluded to by Ivan Strasky (2002): "They assert that image, the tough-guy image."
72. Ivan Strasky (1998–2002), interview with the author, November 6, 2004.
73. This sentiment was also echoed by an African American colleague who had graduated from high school in the early 1990s, and who said that African Americans no longer considered it acceptable that white researchers interview African Americans. My colleague's position is fully justified given the abusive ways that some white researchers have indeed used "research" to advance their theories of racial superiority, and the biased ways they have developed myriad explanations for minority youth's lower school achievements that bypass or

deny structural forces and institutionalized racism. Her remarks emerged from our discussions around the epistemology of ethnography and more vividly sensitized me to questions of legitimate knowledge-making. Because it is my deep commitment to contribute to scholarly work that helps those whose voices have been bypassed or silenced, and to help historical protagonists "appropriate the social construction of meaning to advance their own interests" (George W. Noblit et al., 2004, 14), I am called to be transparent. I am fully aware that to participate in exploring intergroup relationships through oral histories, I am to make transparent my middle-class whiteness so that whatever racial, class, or gender bias might infuse my work against my best efforts to remain as honest a researcher as possible, my colleagues—across racial, gender, class, or other divides—may more readily identify them as such. Oral historical research that explores multiple identities and border crossings among and between very diverse groups of people will always present the challenge of identity encounters between interviewer and interviewee. I am hopeful that as oral historians of education continue to make their researcher identities transparent, more honest articulations of our human investigations will emerge. For a thoughtful discussion of historical evolution of ethnographic inquiry in education and policy, and implications for emancipation of underrepresented voices, see George W. Noblit, Susana Y. Flores, and Enrique G. Murillon, Jr., *Postcritical Ethnography* (Cresskill, NJ: Hampton Press, 2004).
74. Ivan Strasky (1998–2002), interview with the author, November 6, 2004.
75. Ibid.
76. David Randle, a 1976 African American graduate, said that as a teacher and coach throughout the 1990s, he was more frequently breaking up fights between girls: "Now...fights every day, mostly girls."
77. Betty Ames (1995–1999), interview with the author, August 19, 2003.

6 Oppositional Self-Segregation: A Student Body Sensitized to Discrimination (1986–2000)

1. Vera Debin (1995–1999), interview with the author, January 15, 2004.
2. For a critical examination of the political power of elite parents to shape school curricula, and a discussion of how upper-class parents' ubiquitous presence in school classrooms, on school boards, and in myriad extracurricular activities, in the form of volunteer work and monetary contributions, buy them the political clout to lobby for ever more specialized top-end programs that privilege their children (GT, honors, AP classes), relegating the children of working-class and poor parents to the regular and remedial classes, see Peter Sacks, *Tearing Down the Gates: Confronting the Class Divide in American Education* (Berkeley and Los Angeles California: University of California Press, 2007).

3. Vera Debin (1995–1999), interview with the author, January 15, 2004.
4. Ivan Strasky (1998–2002), interview with the author, November 6, 2004.
5. Refer to testimonies of graduates of the Divided and Border-Crossing Generations in general. Also, in speaking of students who did not belong to peer groups during the 1970s and early 1980s: Joanne Pet (1970–1974) (interview with the author, October 16, 2003) recalled: "You segregated yourself regardless... It's just a natural thing. We do it every day. Subconsciously we do it. So, it's a comfortable type of segregation... Didn't you find that you could walk into the library and you had your Asian people kind of sitting here, black people sitting there... and you might have a Caucasian table here... You know, like on my job, we have maybe four Asian people, and they kind of do sit together in the canteen, they speak their own language... They feel more comfortable."
6. By no means do I suggest, as have historian Arthur Schlesinger and others, who believe that affirmations of difference ultimately further marginalize the already marginalized and "disunite America," that multicultural education in itself leads to fragmentation and division among students. However, findings in this history alert us to consider multicultural education through international perspectives, and further, to consider the role of broader institutional norms (such as ever-escalating tracking, uninvolved and distant faculty, etc.) in counteracting efforts to foster respect for difference through the curriculum.
7. Tyack and Cuban, *Tinkering Toward Utopia: A Century of Public School Reform* (Cambridge, MA: Harvard University Press, 1995), 44–45.
8. This finding was further corroborated by the testimony of Sam Buck, a Spanish teacher with more than twenty-five years of tenure who reflected on Miller High in the 1990s. Sam Buck, (interview with the author, November 7, 2003): "you had two parallel worlds"—the world of those whom Sam called the "mainstream" students, and those he placed in the category of "antisocial students." He said that students mostly went about their business—the mainstream not being bothered by the "troublemaker" constituency, and the troublemaker constituency not interested in the "mainstream" world.
9. Michelle Fine, Lois Weiss, and Linda C. Powell, *Harvard Educational Review* 67 (1997): 248.
10. See William Wraga, *Democracy's High School: The Comprehensive High School and Educational Reform in the United States* (Lanham, MD: University Press of America, 1994); see also Michael Sedlack, Christopher Wheeler, Diana C. Pullin, and Philip A. Cusick, *Selling Students Short: Classroom Bargains and Academic Reform in the American High School* (NY: Teachers' College Columbia University, 1986).

Conclusion

1. This finding is confirmed to some extent by research conducted by Amy Stuart Wells et al.: "We found more cross-racial friendships in high schools

with larger middle-class Black student populations." See "How Society Failed School Desegregation Policy: Looking Past the Schools to Understand Them," *Review of Research in Education*, 28 (2005): 83.
2. Findings by Amy Stuart Wells et al. also suggest "that there are several ways to interpret lunchroom segregation within desegregated schools...and one is to consider who students see and interact with outside of school." Wells (2005), 90. Certainly there are indications that even at the beginning of desegregation at Miller High, during the Divided Generation, interracial playmates of Miller Town were more likely to sit together in the cafeteria during lunchtime. Refer to Annie Cole's testimonies. I suggest that beyond imported habits of association, students also segregated in the cafeteria at the end of the century because of the absence of adult role-modeling and adult engagement.
3. Wells et al., "How Society Failed School Desegregation Policy: Looking Past the Schools to Understand Them," 90: "Close friendships within racially diverse schools are more often than not same-race friendships...." This historical analysis shows, however, that interracial friendships can flourish under particular social and demographic conditions such as those experienced by the Border-Crossing Generation.
4. Recent studies have documented patterns of school resegregation across the nation and warned of the dangers to a diverse democratic society of public practices that racially divide its youth. However, while these studies question the fate of the civil rights movements, they build arguments on the assumption that desegregated schools, by virtue of being desegregated, foster integrated environments. See report by Gary Orfield and Chungmei Lee, "Brown at Fifty: King's Dream or Plessy's Nightmare," Civil Rights Project Harvard University, http://www.gse.harvard.edu/news_events/features/2004/orfield01182004.html (accessed July 2, 2010).
5. Amy Stuart Wells, Jennifer Jellison Holme, Anita Tijerina Revilla, and Awo Korantemaa Atanda, "How Desegregation Changed Us: The Effects of Racially Mixed Schools on Students and Society," http://faculty.tc.columbia.edu/upload/asw86/ASWells041504.pdf (accessed July, 2009). On page five of the document, the authors state that they interviewed "more than 500 graduates, educators, advocates, and local policy makers who were direct-lyinvolved in racially mixed public high schools in different communities 25 years ago."
6. The state of Maryland, along with most other U.S. states in the 1990s, made changes to its laws governing juvenile justice and held more young offenders accountable through adult sentencing. See Megan C. Kurlycheck and Brian D. Johnson, "The Juvenile Penalty: A Comparison of Juvenile and Young Adult Sentencing Outcomes in Criminal Court," *Criminology* 42 (2004): 485–515. Wiley InterScience.
7. For examples of histories of education that address the high school's sorting mechanisms, see works by Samuel Bowles and Herbert Gintis, *Schooling in Capitalist America* (NY: Basic Books, 1976); James Anderson, *The Education of Blacks in the South, 1860–1935* (Chapel Hill, NC: University of North

Carolina Press, 1988); and William Reece, *The Origins of the American High School* (New Haven, CT: Yale University Press, 1995). According to Bowles and Gintis, school hierarchy of power and authority parallels the organization of power and authority in the workplace. The authors liken the role of grades to that of wages and establish a direct correspondence between competition among students and their lack of control over the curriculum with competition among workers and their lack of control over required contents of their assigned tasks. Bowles and Gintis advance that the American educational system serves the purposes and needs of the "production process and structure of class relations in the United States." In *The Education of Blacks in the South, 1860–1935*, Anderson identifies dominant paradigms of social reproduction in the Hampton Model of Normal School Industrial Education for blacks (Chapter 2) and tracks reproduction of castes in the black public high school of the South (Chapter 6). For reflections on the sorting mechanisms of public schools in general, see works by David Tyack, *The One Best System: A History of American Urban Education* (Cambridge, MA: Harvard University Press, 1974); David Tyack and Elizabeth Hansot, *Managers of Virtue: Public School Leadership in America, 1820–1980* (NY: Basic Books, 1982); and Michael Katz, *The Irony of Early School Reform* (NY: Teachers College Press, 1968). These historians of education identify the modernizing tendency of schools to prepare youth for various workstations in society, to assure that youth is "properly socialized to the new modes of production, attuned to hierarchy, affective neutrality, role-specific demands, extrinsic incentives for achievement" (Tyack, 1974, p. 73). For ethnographies that focus on the high school and expose its divisive and exclusionary practices, see Michelle Fine, *Framing Dropouts* (Albany: State University of New York Press, 1991); Ellen Bratlinger, *The Politics of Class in Secondary School* (NY: Teachers College Press, 1993); Jay MacLeod, *Ain't No Makin' It* (Boulder, CO: Westview Press, 1995; Anne Locke-Davidson, *Making and Molding Identity in School* (NY: SUNY Press, 1996). See also Jeanie Oakes, *Keeping Track: How Schools Structure Inequality* (New Haven, CT: Yale University Press, 2006); Jonathan Kozol, *Savage Inequalities: Children in America's Schools* (NY: Harper Collins, 1991). These ethnographies expose institutionalized racism and classism.

8. Anthropologist John Ogbu attributes student behavior within school to cultural influences outside of school. By comparing involuntary minorities' cultural backgrounds, Ogbu concludes that involuntary minorities (e.g., black youth) inherit through "family and community discussions and gossip, as well as through public debates over minority education" mistrust of school and school authorities and a deeply ingrained skepticism about their chances to succeed in schools. His work "Voluntary and Involuntary Minorities: A Cultural-Ecological Theory of School Performance with Some Implications for Education," *Anthropology and Education Quarterly* 29 (1998): 155–188, has received serious criticism. My own analysis challenges the notion that immigrants do not share with those whom Ogbu calls "involuntary minorities" the same distrust of school authorities, as

evidenced in voluntary Russian immigrant students' mistrust and even disdain of American teachers. Historians of education, as explained in the Introduction, have barely examined the high school from students' perspectives.

9. Schofield (1989), quoted in Amy Stuart et al., "How Society Failed School Desegregation Policy: Looking Past the Schools to Understand Them," *Review of Research in Education* 28 (2005): 71.
10. Jencks, quoted in Amy Stuart Wells et al., "How Society Failed School Desegregation Policy: Looking Past the Schools to Understand Them," *Review of Research in Education* 28 (2005): 71.
11. Gerald Grand, *The World We Created at Hamilton High*, 210.
12. Patrick Ryan, "A Case Study in the Cultural Origins of a Superpower: Liberal Individualism, American Nationalism, and the Rise of High School Life, A Study of Cleveland's Central and East Technical High Schools, 1890–1918," *History of Education Quarterly*, 45 (Spring 2005): 67.
13. Patrick Ryan, "A Case Study in the Cultural Origins of a Superpower: Liberalism, Individualism, American Nationalism, and the Rise of High School Life, A Study of Cleveland's Central and East Technical High Schools," *History of Education Quarterly*, 45 (Spring 2005): 94.

Addendum

1. See Donald Ritchie, *Doing Oral History* (New York: Twayne Publishers, 1995).
2. Alessandro Portelli states: "[T]he first thing that makes oral history different, is that it tells us less about *events* than about their *meaning*... what informants believe is indeed a historical *fact* (that is, the fact that they believe it), as much as what really happened." Alessandro Portelli, "What makes oral history different," in *The Oral History Reader*, Robert Perks and Alistair Thomson, eds. (New York: Routledge, 2005), 63–74. See also Alessandro Portelli, *The Death of Luigi Trastulli and Other Stories: Form and Meaning in Oral History* (Albany: SUNY Press, 1991).
3. See Penny Summerfield, *Reconstructing Women's Wartime Lives: Discourse and Subjectivity in Oral Histories of the Second World War* (Manchester: Manchester University Press, 1998); Alessandro Portelli, *The Death of Luigi Trastulli and Other Stories: Form and Meaning in Oral History* (Albany: SUNY Press, 1991).
4. See "Ethnicity and National Identity," special issue, *Oral History* 21 (1993); Susan K. Burton, "Issues in Cross-Cultural Interviewing: Japanese Women in England," *Oral History* 31 1(2003); Jelena Cvorovic, "Gypsy Oral History in Serbia: From Poverty to Culture," *Oral History* 33 1 (2005); "Black History," special issue, *Oral History* 8 1(1980).
5. Paul Thompson, *The Voice of the Past: Oral History* (Oxford: 1978), 7–8.

6. Ronald J. Grele, "Movement without Aim: Methodological and Theoretical Problems in Oral History," in *The Oral History Reader*, Robert Perks and Alistair Thomson, eds. (New York: Routledge, 2005), 38–53.
7. Mikhail Bakhtin, quoted in Donald Freeman, "To Take Them at Their Word: Language Data in the Study of Teachers' Knowledge," *Harvard Educational Review* 66 (1996): 748–750.
8. See Patricia Collins, *Black Feminist Thought: Knowledge, Counsciousness, and the Politics of Empowerment* (New York: Routledge, 2000); Patricia Collins, *Fighting Words: Black Women and the Search for Justice* (Minneapolis: University of Minnesota Press, 1998).
9. Leslie McCall, "The Complexity of Intersectionality," *Signs: Journal of Women in Culture and Society* 30 (2005): 1771.
10. Leslie McCall, "The Complexity of Intersectionality," *Signs: Journal of Women in Culture and Society* 30 (2005): 1771–1800.
11. "Intersectionality," http://en.wikipedia.org/wiki/Intersectionality (accessed June 15, 2009).
12. One of many examples of documentary data that contextualized oral testimonies involved the Russian middle-class alumna Vera Debin, a member of the Redivided Generation (1986–2000). Debin shared her feelings about the yearbook pictures—"they [the preps] went around and tried to get pictures of all the clubs, and all the band, but it was pretty much them and their friends, all extra pictures"—and further suggested that yearbooks would "probably be more interesting if people were assigned to a committee to do the yearbook, randomly selected" to ensure broad and fair representations of all students. I subsequently examined yearbooks for redundancy in pictures. Indeed, the same faces would reappear on the pages of the yearbooks. When graduates recalled what kind of student attended what kind of academic track, extracurricular activity, or sport, I systematically analyzed yearbooks for the distribution of students across extracurricular activities and curricular tracks, by race, ethnicity, and gender.
13. Irma Omeldo, "Redefining Culture through the Memorias of Elderly Latinas," *Qualitative Inquiry* 5 (1993): 353–376.
14. Constructing minimum or set numbers of testimonies becomes even more problematic when one considers the disparity in numbers between white and black students in the early stages of desegregation. For example, how many of the total 121 white students who graduated in 1958 should one interview to "counterbalance" interviews with the four African American students who graduated the same year? In relational histories of education, "difference" becomes a more relevant and workable criterion of selection than numbers and volume.
15. See works by Louis S. Diggs, *Since the Beginning: African American Communities in Towson* (Baltimore: Uptown Press, 2000); *Holding on to Their Heritage* (Catonsville, MD: privately printed, 1996); *In Our Own Voices: A Folk History in Legacy* (Catonsville, MD: privately printed, 1998).
16. I have since discovered the NviVo coding program, which I consider valuable for coding interviews.

17. See Paul Thompson, *The Voice of the Past: Oral History* (Oxford: Oxford University Press, 1988), 131.
18. Paula Fass, *Generations of Youth: Youth Culture and History in Twentieth-Century America*, 95–117.
19. Michael Frisch, *A Shared Authority: Essays on the Craft and Meaning of Oral and Public History* (New York, 1990), xx.

Bibliography

Anderson, James. *The Education of Blacks in the South, 1860–1935.* Chapel Hill, N.C.: University of North Carolina Press, 1988.
Anderson, James and Dara Byrne, eds. *The Unfinished Agenda of Brown v. Board of Education.* Hoboken, NJ: John Wiley & Sons, 2004.
Ashcroft, Bill, Gareth Griffiths, and Helen Tiffin, eds. *Post-Colonial Studies: The Key Concepts.* NY: Routledge, 2005
Austin, Joe, and Michael Willard, eds. *Generations of Youth: Youth Culture and History in Twentieth Century America.* NY: New York University Press, 1998.
Bailey, Beth. "From Panty Raids to Revolution: Youth and Authority, 1950–1970." In *Generations of Youth: Youth Culture and History in Twentieth Century America*, edited by Joe Austin and Michael Willard, 187–204. NY: New York University Press, 1998.
———. *From Front Porch to Back Seat: Courtship in Twentieth Century America* Baltimore: The Johns Hopkins University Press, 1998.
Black Grooves. Afro-Punk. Archives of American Music and Culture. http://blackgrooves.org/?p=283 (accessed July 17, 2009).
Boli, John, Francisco O. Ramirez, and John W. Meyer, "Explaining the Origins of Expansion of Mass Education." *Comparative Education Review* 29 (1985): 145–170.
Bowles, Samuel, and Herbert Gintis. *Schooling in Capitalist America.* NY: Basic Books, 1976.
Bratlinger, Ellen, A. *The Politics of Class in Secondary School.* NY: Teachers College Press, 1993.
Brown, Roger. *Social Psychology.* NY: Free Press, 1986.
Burke, Peter. *History and Social Theory.* NY: Cornell University Press, 2005.
Clayborne Carson, "Two Cheers for Brown v Board of Education," *Journal of American History* 91 (2004): 26–31.
Chism, Khalil. "A Documentary History of Brown." In *The Unfinished Agenda of Brown v. Board of Education*, edited by James Anderson and Dara Byrne, 7–22. Hoboken, New Jersey: John Wiley & Sons, 2004.
Cohen, Ronald. "The Delinquents: Censorship and Youth Culture in Recent U.S. History." *History of Education Quarterly* 37, no. 3 (1997): 251–270.
Collins, Patricia. *Black Feminist Thought: Knowledge, Consciousness, and the Politics of Empowerment.* NY: Routledge, 2000.

Collins, Patricia. *Fighting Words: Black Women and the Search for Justice.* Minneapolis: University of Minnesota Press, 1998.
Coleman, James. *The Adolescent Society: The Social Life of the Teenager and its Impact on Education.* NY: The Free Press of Glencoe, 1961.
Conant, James, B. *The American High School Today: A First Report to Interested Citizens.* NY: McGraw-Hill, 1959.
Coontz, Stephanie. *The Way We Never Were: American Families and the Nostalgia Trap.* NY: Basic Books, 2000.
Crenshaw, Kimberlé, Neil Gotanda, Gary Peller, and Kendall Thomas, eds. *Critical Race Theory: The Key Writings That Formed the Movement.* NY: New York Press, 1995.
Danns, Dionne. "Chicago High School Students' Movement for Quality Education, 1966–1971." *The Journal of African-American History* 88 (Spring 2003): 138–150.
Davidson, Locke Anne. *Making and Molding Identity in Schools.* NY: State University of New York Press, 1996.
DeHar, Hu. "An Asian American Perspective on Brown." In *The Unfinished Agenda of Brown v. Board of Education*, edited by James Anderson and Dara Byrne, 108–119. Hoboken, NJ: John Wiley & Sons, 2004.
Diggs, Louis. *Since the Beginning: African American Communities in Towson.* Baltimore: Uptown Press, 2000.
———. *In Our Own Voices: A Folk History in Legacy.* Catonsville, MD: Privately Printed, 1998.
———. *Holding On To Their Heritage.* Catonsville, MD: Privately Printed, 1996.
Dorn, Sherman. "Origins of the "Dropout Problem," *History of Education Quarterly* 33, no. 3 (1993): 353–373.
Eckart, Gabriele. "The German Gothic Subculture." *German Studies Review* 28 (2005): 547–562.
Eckert, Penelope. *Jocks & Burnouts: Social Categories and Identity in the High School.* NY: Teachers College Press,1989.
Eisenman, Linda. *Higher Education for Women in Postwar America, 1945–1965: Reclaiming the Incidental Student.* Baltimore: Johns Hopkins University Press, 2006.
Eisner, Elliot, W. and Alan Peshkin, eds. *Qualitative Inquiry in Education: The Continuing Debate.* NY: Teachers College, 1990.
Fass, Paula. "Creating New Identities: Youth and Ethnicity in New York City High Schools in the 1930s and 1940s." In *Generations of Youth: Youth Culture and History in Twentieth Century America*, edited by Joe Austin and Michael Willard, 95–117. NY: New York University Press, 1998.
———. *The Damned and the Beautiful: American Youth in the 1920s.* NY: Oxford University Press, 1977.
Fine, Michelle. *Framing Dropouts.* Albany: State University of New York Press, 1991.
Fine, Michelle, Lois Weiss, and Linda C. Powell. "Communities of Difference: A Critical Look at Desegregated Spaces Created by and for Youth". *Harvard Educational Review* 67 (Summer 1997): 247–285.

Finkelstein, Barbara. "Is Adolescence Here to Stay?: Historical Perspectives on Youth and Education." In *Adolescence and Society*, edited by Timothy C. Urdan and Frank Pajares, 1–33. NY: Information Age Press, 2003.

Fish, Kenneth. *Conflict and Dissent in the High School*. NY: Bruce Publishing Company, 1970.

Franklin, V.P. "Black High School Activism in the 1960s: An Urban Phenomenon?" *Journal of Research in Education*. 10, no.1 (2000): 3–8.

Frazier, E.F. *Negro Youth at the Crossroads: Their Personality Development in the Middle States*. NY: Scholar Books, 1967.

Freeman, Donald. "'To Take Them at Their World': Language Data in the Study of Teachers' Knowledge." *Harvard Educational Review* 66 (Winter 1996): 732–761.

Frisch, Michael. *A Shared Authority*. Albany: State University of New York Press, 1990.

Gilbert, John. *Cycles of Outrage: America's Reaction to the Juvenile Delinquent in the 1950s*. NY: Oxford University Press, 1986.

Gillis, John. *Youth and History: Tradition and Change in European Age Relations 1779-Present*. NY and London: Academic Press, Inc., 1974.

Grant, Gerald. *The World We Created at Hamilton High*. Cambridge: Massachusetts: Harvard University Press, 1988.

Graebner, William. *Coming of Age in Buffalo: Youth and Authority in the Postwar Era*. Philadelphia: Temple University Press, 1990.

Grele, Ronald, J. "Movement without Aim: Methodological and Theoretical Problems in Oral History." In *The Oral History Reader*, edited by Robert Perks and Alistair Thomson, 38–53. NY: Routledge, 2005.

Grundy, Pamela. *Learning to Win: Sports, Education, and Social Change in Twentieth Century North Carolina*. Chapel Hill: University of North Carolina, 2001.

Gutmann, Amy. *Democratic Education*. Princeton, NJ: Princeton University Press, 1999.

Hall, Stanley. "How Far Is the Present High School and Early College Training Adapted to the Nature and Needs of Adolescents?" *School Review* 9 (1901): 649–681.

Hollingstead, August, B. *Elmtown's Youth: The Impact of Social Class and Adolescents*. NY: John Wiley & Sons, 1949.

Irons, Peter. *Jim Crow's Children: The Broken Promise of the Brown Decision*. New York, NY: Viking, 2002.

Jackson, Kenneth, T. *Crabgrass Frontier: The Suburbanization of the United States*. Oxford, NY, Toronto: Oxford University Press, 1985.

Katz, Michael. *Class, Bureaucracy, and the Schools*. NY: Praeger, 1971.

———. *The Irony of Early School Reform*. NY: Teachers College, 1968.

Kett, Joseph. *Rites of Passage: Adolescence in America, 1790 to the Present*. New York: Basic Books, 1977.

Kozol, Jonathan. *Savage Inequalities: Children in America's Schools*. NY, Harper Collins, 1991.

Kurlycheck, Megan, C. and Brian D. Johnson. "The Juvenile Penalty: A Comparison of Juvenile and Young Adult Sentencing Outcomes in Criminal

Court." Criminology 42 (2006): 485–515, http://www3.interscience.wiley.com/journal/118749263/abstract?CRETRY=1&SRETRY=0

Labaree, David. *The Making of an American High School: The Credentials Market and the Central High School of Philadelphia, 1838–1939*. New Haven: Yale University Press, 1988.

———. *How to Succeed in School Without Really Learning*. New Haven: Yale University Press, 1997.

Lemann, Nicholas. *The Promised Land: The Great Black Migration and How It Changed America*. NY: Alfred A. Knopf, Inc., 1991.

Levine, Robert and Donald Campbell. *Ethnocentrism: Theories of Conflict, Ethnic Attitudes, and Group Behavior*. NY: John Wiley & Sons, 1972.

Levesque, Roger, JR. *Dangerous Adolescents, Model Adolescents: Shaping the Role and Promise of Education*. NY: Kluwer Academic/Plenum Publishers, 2002.

Lynd, Robert and Helen Lynd. *Middletown*. NY: Harcourt Brace, 1929.

MacDonald, Victoria-María, ed. *Latino Education in the U.S.: A Narrative History, 1531–2000*. NY: Palgrave/Macmillan, 2004.

MacLeod, Jay. *Ain't no Makin' It*. Boulder, CO: Westview Press, 1995.

McCall, Leslie. "The Complexity of Intersectionality." Signs: Journal of Women in Culture and Society 30 (2005): 1771–1800, http://www.journals.uchicago.edu/doi/abs/10.1086/426800

Marginson, Simon. "After Globalization, Emerging Politics of Education." *Journal of Education Policy* 14 (1999): 19–35.

Modell, John. *Into One's Own: From Youth to Adulthood in the United States, 1920–1975*. Berkeley: University of California Press, 1989.

Noblit, George, Susana Y. Flores, and Enrique G. Murillon, Jr. *Postcritical Ethnography*. Cresskill, NJ: Hampton Press, Inc., 2004.

Oakes, Jeanie. *Keeping Track: How Schools Structure Inequality*. New Haven, CT: Yale University Press, 2006.

Olsen, Laurie. *Made in America: Immigrant Students in our Public Schools*. NY: The New Press, 1997.

Ogbu, John. "Voluntary and Involuntary Minorities: A Cultural-Ecological Theory of School Performance with Some Implications for Education." *Anthropology and Education Quarterly* 29 no. 2 (1998): 155–188.

Omeldo, Irma. "Redefining Culture through the Memorias of Elderly Latinas." *Qualitative Inquiry* 5 (1993): 353–376.

Orfield, Gary and Chungmei Lee. "Brown at Fifty: King's Dream or Plessy's Nightmare." The Civil Rights Project Harvard University. (Jan. 17, 2004), http://www.%20.civilrightsproject.harvard.edu/research/reseg40/resegregation04.php

Orfield, Gary. "Schools More Separate: Consequences of a Decade of Resegregation." Harvard University, The Civil Rights Project Web site, http://www.civilrightsproject.harvard.edu/research/deseg/separate_schools01.php

Palladino, Grace. *Teenagers: An American History*. Chicago: University of Chicago Press, 1996.

Palumbos, Robert. "Student Involvement in the Baltimore Civil Rights Movement, 1953–63." *Maryland Historical Magazine* 94 (1999): 449–492.

Peshkin, Alan. *Places of Memory: Whiteman's Schools and Native American Communities.* Mahwah, NJ: Lawrence Erlbaum Associates, 1997.

Portales, Marco. "A History of Latino Segregation Lawsuits." In *The Unfinished Agenda of Brown v. Board of Education*, edited by James Anderson and Dara Byrne, 123–136. Hoboken, NJ: John Wiley & Sons, Inc., 2004.

Portelli, Alessandro. "What Makes Oral History Different." In *The Oral History Reader*, edited by Robert Perks and Alistair Thomson, 63–74. NY: Routledge, 2005.

———. *The Death of Luigi Trastulli and Other Stories: Form and Meaning in Oral History.* Albany: SUNY Press, 1990.

Portes, Alejandro and Rubén G. Rumbaut. *Immigrant America: A Portrait.* Berkeley and Los Angeles: The University of California Press, 1996.

Powell, Arthur, Eleanor Farrar, and David Cohen. *The Shopping Mall High School.* Boston: Houghton Mcfflin Co., 1985.

Price, Jeremy. *Against the Odds: The Meanings of School and Relationships in the Lives of Six Young African-American Men.* Stamford, Connecticut: Alex Publishing Corporation, 2000.

Reese, William. *The Origins of the American High School.* New Haven: Yale University Press, 1995.

Ritchie, Donald. *Doing Oral History.* NY: Twayne Publishers, 1995.

Ryan, Patrick J. "A Case Study in the Cultural Origins of a Superpower: Liberal Individualism, American Nationalism, and the Rise of High School Life, A Study of Cleveland's Central and East Technical Schools, 1890–1918." *History of Education Quarterly* 45 (Spring 2005): 66–95.

Sacks, Peter. *Tearing Down the Gates: Confronting the Class Divide in American Education.* Berkeley and Los Angeles California: University of California Press, 2007.

Salomone, Rosemary, C. *Vision of Schooling: Conscience, Community and Common Education.* New Haven: Yale, 2000.

Sedlack, Michael, et al. *Selling Students Short: Classroom Bargains and Academic Reform in the American High School.* NY: Teachers' College Columbia University, 1986.

Shlossman, Stephen. *Love and the American Delinquent.* Chicago: University of Chicago Press, 1977.

Summerfield, Penny. *Reconstructing Women's Wartime Lives: Discourse and Subjectivity in Oral Histories of the Second World War.* Manchester: Manchester University Press, 1998.

Talburt, Susan. "Time, Space, and Ethnography Without Proper Subjects." In *Postcritical Ethnography*, edited by George Noblit, Susana Flores, and Enrique Murillo, 107–123. Cresskill, NJ: Hampton Press, Inc., 2004.

Thompson, Paul. *The Voice of the Past: Oral History.* Oxford: Oxford University Press, 1988.

Thompson, Stacey. "Punk Cinema." *Cinema Journal* 43 no.2 (2004): 47–66, http://muse.jhu.edu/login?uri=/journals/cinema_journal/v043/43.2thompson.html

Tyack, David and Larry Cuban. *Tinkering Toward Utopia: A Century of Public School Reform.* Cambridge, Massachusetts: Harvard University Press, 1995.

Tyack, David and Elizabeth Hansot. *Managers of Virtue: Public School Leadership in America, 1820–1980.* NY: Basic Books, 1982.
Urban, Wayne and Jennings Wagoner. *American Education.* NY: McGraw Hill Companies, Inc., 2000.
Vicino, Thomas, J. *Transforming Race and Class in Suburbia: Decline in Metropolitan Baltimore.* NY: Palgrave Macmillan, 2008.
Wells Stuart, Amy, et.al. "How Society Failed School Desegregation Policy: Looking Past the Schools to Understand Them." *Review of Research in Education* 28 (2005): 47–99.
———. "How Desegregation Changed Us: The Effects of Racially Mixed Schools on Students and Society." http://faculty.tc.columbia.edu/upload/asw86/ASWells041504.pdf
West, Elliot. *Growing Up In Twentieth Century America: A History and Reference Guide.* Westport, Connecticut: Greenwood Press, 1996.
Willis, Paul. *Learning to Labor.* NY: Columbia Press, 1977.
Wraga, William. *Democracy's High School: The Comprehensive High School and Educational Reform in the United States.* Lanham, Maryland: University Press of America, 1994.
Wright, Dwayne, C. "Black Pride Day, 1968: High School Student Activism in York, Pennsylvania." *The Journal of African-American History* 88 (Spring 2003): 151–162.
Zinn, Howard. *A People's History of the United States.* NY: HarperCollins Inc., 2003.

Index

academic achievement, 2, 33, 54, 152
administrators, 6, 17, 18, 23, 29, 48, 94, 95, 105, 136, 138, 139, 142, 202
adolescence, 8, 117, 178, 193, 194, 195
African Americans, xviii, 3, 4, 5, 11, 17, 18, 27, 30, 32, 33–65, 72–98, 122–144, 151, 153, 161, 162, 164, 168, 169, 171, 173, 174, 179, 180, 183, 190, 194, 195, 197, 199, 200, 201, 202, 206, 207, 212
see also blacks
assimilation, 12, 176, 177
associative living, xviii

Bakhtin, Mikhail, 158, 212
Baltimore city, 3, 172, 174
Baltimore County, xii, xvii, 1, 3, 17, 18, 30, 35, 165, 172, 190
Bilingual Education Act, xvi
blacks, xii, xiii, xv, xviii, 1, 3–6, 9–12, 13, 18, 19, 26, 27, 30, 31–68, 73, 90–110, 115, 116, 121–137, 138–140, 143–150, 151–155, 159, 161–164, 166, 170, 171, 173, 174, 175, 177, 179, 183, 190, 191, 192, 193, 194, 196, 200, 201, 203, 204, 205, 206, 209, 210, 211, 212
black, middle-class, 10, 31, 80, 150
black, women, 33, 34, 38, 39, 40, 47, 51–55, 88, 89, 97, 150, 151, 190
black feminist theory, 6, 159, 180, 212, 215

black men, 35, 36, 38, 40, 41, 46, 48, 49, 53, 65, 68, 82, 97, 98, 150, 173
see also African Americans
Brown V. Board of Education, xii, xv, 8, 9, 145, 175, 182

Cafeteria, 10, 11, 32, 52, 57, 64, 78, 87, 88, 89, 110, 111, 122, 130, 139, 142, 144, 148, 150, 151, 152, 182, 209
Carnoy, Martin and Levin, Henry, 9, 57, 142, 150, 182
Christian, 37, 49, 50, 105, 107, 109, 110, 113, 122, 126, 133, 134, 142, 145
city, xiii, xvi, xvii, 2–5, 8, 9, 10, 18, 40, 62, 98, 103, 104, 124, 128, 129, 130, 135, 136, 137, 139, 145, 148, 153, 172, 178, 185, 201
civil rights, xii, xv, xvii, 3, 4, 5, 8, 10, 87, 88, 147, 149, 154, 175, 176, 179, 182, 209, 218
class (economic status), xii, xiii, xv, xvi, xviii, 1, 2, 4–13, 17–90, 95–97, 99, 105, 108, 114, 117, 119, 121, 122, 126, 131, 133, 135, 137, 138, 139, 140, 145, 147, 148, 150, 151, 152, 153, 155, 157–161, 166, 167, 168, 170, 178, 179, 180, 183, 188, 193, 207, 209, 210, 217
lower-class, 10, 21, 24, 27, 28, 38, 43, 46, 56, 84, 85, 96, 97, 133, 139, 147, 166, 167, 168, 169, 207, *see also* working-class students

class (economic status)—*Continued*
  middle-class, 4, 10, 12, 17–57, 66, 71, 72, 78, 80, 82, 83, 84, 90, 107, 108, 111, 114, 117, 119, 122, 127, 133, 138, 139, 140, 147, 148, 150, 153, 166, 168, 169, 183, 193, 207, 209
classroom, xviii, 19, 20, 22, 28, 31, 36, 46, 50, 72, 77, 80, 94, 98, 114, 151, 187, 202, 207, 208
coding, 157, 170, 212
Coleman, James, 55, 175
Collins, Patricia, 6, 159, 177
community, 4, 5, 10, 11, 17, 19, 21, 24, 25, 28, 33, 36, 38, 48, 49, 52, 76, 96, 98, 103, 106, 107, 112, 117, 118, 119, 128, 129, 130, 139, 142, 147, 148, 150, 151, 153, 158, 160, 161–163, 165, 172, 177, 182, 184, 189, 191, 195, 201, 203, 210
comprehensive public school, xvi, 154
conservative students / town, xvii, 5, 10, 18, 45, 47–57, 126, 134, 138, 142, 143, 146, 147, 148, 181, 192
consumerism, 126, 131, 145
cross-group relationships, 1, 6, 7, 9, 10, 159, 160, 161, 171
  cross-class relationships, 2, 4, 5, 8, 10, 13, 52, 76, 138, 148
  cross-racial relationships, 52, 76, 109, 150, 199, 208
culture, xv, 9, 10, 13, 17, 23, 25, 28, 43, 72, 78, 97, 113, 114, 122, 129, 136, 148, 151–152, 153, 154, 158, 159, 160, 171, 178, 192, 195, 208, 210

dances, xviii, 19, 22, 38, 39, 41, 42, 49, 50, 85, 90, 108, 121, 151, 193
dating, 22, 23, 49, 50, 56, 91, 185
democracy, xv, xvi, xviii, 1, 20, 154, 155, 181
demographics, xviii, 4, 11, 18, 83, 103, 133, 134, 135, 148, 152, 153

desegregation, xii, xiii, xv, xvii, xviii, 1, 2, 3, 11, 12, 17, 18, 19, 30, 33, 34, 36, 37, 38, 39, 41, 42, 43, 48, 50, 51, 52, 56, 62, 74, 96, 107, 125, 138, 143, 144, 149, 150, 153, 158, 164, 167, 168, 175, 177, 178
diploma, 20, 23, 27, 33, 37, 55, 75, 81, 82, 118, 165, 185
discipline, 48, 70, 78, 94, 106, 108, 138, 196
discrimination, 41, 51, 76, 80, 88, 113, 125, 128, 142, 143, 144, 150, 151, 153

economic status, xv, xvi, 11, 18, 29, 42, 43, 44, 48, 52–55, 62, 70, 83, 84, 90, 99, 105, 113, 114, 117, 119, 121, 122, 126, 128, 138, 148, 151, 152, 154, 155, 159–163, 166, 176
Eisenman, Linda, 50, 183
Elementary and Secondary Education Act, xvi
ethnicity, xv, xvi, xviii, 1, 6, 12, 122, 133, 136, 137, 152, 154, 159, 161, 173
expulsions, 105, 106, 122, 141, 145
extracurricular activities, xvi, xviii, 10, 18, 21, 22, 40, 46, 47, 52, 67, 68, 80, 83, 98, 99, 108, 112, 119, 123, 129, 135, 147, 148, 173, 178, 192, 195, 203, 207

family, 19, 21, 22, 23, 34, 36, 43, 46, 48, 49, 51, 62, 69, 71, 72, 76, 80, 81, 166, 168, 176, 184, 191, 192, 194, 203, 210
farmers, 18, 25, 62, 116, 128, 135, 138, 139, 173, 183
Fass, Paula, 173, 179, 203

gender, xii, xvi, xvii, xviii, 1, 2, 5, 6, 7, 8, 9, 11, 13, 17, 43, 44, 45, 49, 50, 52, 53, 56, 57, 76, 86, 87,

88, 89, 90, 91, 92, 137, 143, 144, 152, 157, 158, 159, 161, 168, 170, 171, 173, 178, 197, 203
generations, xii, xiii, xv, xvii, xviii, 1, 9, 10, 18, 19, 39, 55, 56, 61, 62, 68, 70, 72, 76, 79, 89, 91, 106, 114, 117, 120, 121, 130, 133, 135, 136, 139, 140, 142, 143, 144, 145, 147, 148, 158, 160, 167, 168, 170, 171, 178, 179, 203
Gerald Grant, 2, 4, 149, 152, 178
globalization, xiii, xvi, xvii, 154, 176

hallways, xviii, 31, 51, 74, 92, 105, 106, 114, 141, 202
historical consciousness, 157, 159, 160, 161

identity, 6, 8, 9, 43, 45, 52, 57, 88, 98, 105, 110, 111, 112, 117, 121, 122, 130, 131, 133, 136, 143, 145, 161, 177, 178, 180
immigration, xv, xvii, 4, 175, 176, 180
injustice, 5, 64
institution (institutional, institutionalized), 3, 5, 7, 8, 10, 11, 12, 13, 17, 19, 21, 30, 37, 43, 44, 45, 78, 95, 97, 98, 111, 121, 136, 137, 138, 140, 141, 144, 147, 151, 152, 153, 154, 157, 158, 160, 161, 162, 163, 164, 171
integration, xviii, 1, 2, 3, 4, 5, 6, 8, 10, 11, 12, 13, 36, 37, 38, 52, 61, 145, 147, 152, 177
interracial relationships, 2, 4, 5, 8, 10, 13, 39, 49, 57, 63, 72, 76, 82, 83, 87, 88, 90, 91, 125, 138, 144, 147, 151, 209
intersectionality theory, 6, 157, 159, 160, 161, 163, 173
interviews, 7, 9, 40, 124, 158, 160, 162, 164, 165, 166, 171, 175, 197, 212

Jewish, 4, 105, 107, 108, 109, 112, 113, 122, 126, 134, 136, 203
jocks, 8, 41, 63, 64, 65, 66, 69–74, 76–79, 82, 87, 89, 90, 91, 92, 95, 86, 99, 107, 109, 111, 126, 134, 151, 180, 192, 196

liberal students (liberalization), xv, xvii, 97, 99, 112, 138, 154, 155

Maryland, xii, xvii, 1, 2, 3, 37, 162, 165, 172, 174, 179, 209
memories, xii, xvii, xix, 1, 2, 9, 10, 13, 17, 19, 24, 28, 30, 40, 61, 64, 65, 71, 79, 91, 92, 103, 106, 108, 116, 117, 126, 127, 130, 140, 157, 158, 160, 165, 170, 171, 182
middle-class, xviii, 4, 10, 12, 17, 18, 19–52, 53, 54, 56, 57, 66, 71, 72, 78, 80, 82, 83, 84, 90, 107, 108, 111, 114, 119, 122, 127, 138, 139, 140, 147, 148, 150, 153, 166, 168, 169, 183, 188, 193, 207, 209, 212
multiculturalism, xv, 176
multiracial schools / demographics, 4, 154, 160

nationality, xi, xii, xv, xvi, xvii, 1, 5, 6, 7, 13, 113, 122, 133, 136, 158, 159, 178
No Child Left Behind Act, xvi

oral history, 7, 14, 157, 159, 180, 201, 211, 212, 213

Palumbos, Robert, 3, 179
peer-groups, 87–199
poor students (economic status), xiii, xviii, 1, 5, 8, 10, 11, 12, 13, 19, 21, 23, 24, 25, 26, 28, 29, 31, 33, 38, 43, 44, 46, 47, 48, 49, 50, 53, 54, 55, 56, 62, 64, 67, 69, 70, 74, 78, 79, 81, 83, 84, 85, 96, 97, 117, 122, 148, 150, 151, 166, 168, 186, 207

Portelli, Alessandro, 7, 180, 211
post-school life opportunities, 2, 53
potheads, 63, 74, 76, 77, 78, 79, 82, 84, 86, 89, 92, 96, 97, 105
pregnant girls, 24, 25, 40, 81, 83, 84, 93, 109, 118, 150, 186
preps, 63, 85, 86, 107, 109, 116, 117, 119, 120, 121–123, 126, 127, 133–135, 138, 139, 151, 154, 168, 169, 205, 212
principal, 19, 21, 22, 23, 30, 32, 33, 38, 41, 57, 70, 92, 94, 95, 96, 106, 107, 114, 116, 124, 127, 141, 144, 166, 187, 191, 199, 204
public schools, xv, xvi, xviii, 1, 2, 4, 18, 46, 55, 107, 136, 154, 180, 181, 209, 210
punks, 8, 105, 109, 126, 127, 145, 206

race, xii, xiii, xv, xvi, xviii, 1, 4, 5, 6, 7, 8, 9, 10, 11, 12, 13, 17, 39, 43, 44, 45, 48, 49, 52, 53, 54, 56, 57, 63, 66, 67, 73, 74, 83, 86, 88, 89, 90, 92, 95, 99, 103, 110, 122, 124, 130, 133, 134, 136–140, 144, 148, 150, 151, 152, 154, 158, 159, 161, 168, 170, 173, 174, 177, 178, 179, 180, 181, 182, 193, 194
racism, 3, 27, 57, 87, 115, 125, 130, 143, 144, 151, 152, 153, 207, 210
religion, 5, 6, 7, 13, 103, 112, 113, 122, 133, 136, 158, 159, 178
resegregation, 1, 2, 5, 99, 176, 182, 209, 218
rich / wealthy students, xviii, 1, 8, 13, 18, 29, 43, 44, 45, 46, 48, 50, 55, 64, 66, 68, 69, 83, 119, 122, 123, 133, 139, 168, 187
 see also wealthy students
rural, xviii, 3, 4, 18, 25, 39, 44, 48, 54, 57, 61, 79, 104, 133, 178, 179, 201
rural-suburban, 4, 133, 178

Russian, xiii, xviii, 4, 5, 10, 103, 104, 105, 109, 110–123, 127, 128, 129–134, 135–143, 145, 148, 151, 153, 154, 163, 165, 169, 174, 201, 204, 205, 206, 211

school spirit, 24, 57, 80, 106, 107, 120, 121, 123, 138, 186, 203
segregation, xii, xv, 1, 2, 3, 5, 8, 11, 13, 52, 53, 56, 62, 74, 89, 96, 97, 103, 126, 131, 133, 134, 142, 143, 148, 149, 151, 152, 182, 208, 209, 219
social capital, 83, 85
social categories, 3, 6, 7, 13, 44, 54, 57, 91, 133, 134, 159, 160, 161, 180
social efficiency, 11, 95, 200
social markers of difference, 6, 159
social memory, 158
social reproduction, 54, 210
sports, xviii, 1, 10, 12, 20, 22, 34, 35, 36, 38, 39, 40, 44, 46, 47, 48, 52, 64, 65, 66, 68, 70, 72, 73, 74, 76, 77, 78, 85, 89, 92, 94, 95, 98, 99, 119, 120, 123, 125, 133, 144, 148, 151, 152, 173, 192, 194, 197, 198, 199, 205, 206
student activism, 3, 5, 179, 200
suburban spaces, xvii, xviii, 2, 3, 4, 8, 61, 62, 78, 79, 83, 88, 104, 114, 126, 127, 133, 178, 183, 206, 217

teachers, 5, 6, 8, 12, 13, 17–29, 31, 32, 34, 38–43, 49, 50, 53–57, 61, 62, 67–72, 75–80, 85, 89–97, 99, 106, 108, 111, 113, 127, 129, 135, 138, 139, 140–143, 150, 151, 153, 165, 166, 173, 176, 183, 184, 185, 187, 188, 189, 191, 193, 194, 196, 197, 200, 203, 204, 206, 211, 212
tracks (tracking), 11, 29, 43, 48, 53, 56, 57, 134, 139, 141, 152, 154, 182, 208

Index

academic track, xv, 1, 10, 11, 12, 21, 22, 28, 29, 33–38, 41, 42, 44, 46–48, 53, 54, 56, 87, 95, 97, 114, 122, 133, 140, 141, 148–154, 173, 190, 203, 212
general track, 26, 28, 29, 38, 42, 43, 44, 48, 52, 53, 113, 122, 185, 194, 200
gifted and talented track (GT), 109, 111, 114, 115, 117, 118, 120, 141, 207
honors track, 108, 109, 114, 115, 120, 203, 207
vocational track, 10, 11, 47, 48, 53, 148, 149, 154
transcriptions, 157, 170

urban spaces, xvii, xviii, 3, 4, 8, 9, 97, 104, 129, 133, 177, 178, 210

violence, 10, 35, 38, 79, 92, 114, 115, 122, 139, 142, 144, 148, 150

wealthy students, xviii, 1, 8, 13, 18, 29, 43, 44, 45, 46, 48, 50, 55, 64, 66, 68, 69, 83, 119, 122, 123, 133, 139, 168, 187
see also rich / wealthy students
whites, xii, xiii, xv, xviii, 1, 3–5, 8–13, 17–63, 65, 66, 67, 69, 70, 71, 72–93, 97, 98, 99, 103, 104, 107–116, 121–128, 133–140, 142–153, 159, 162–164
white, lower-class / working class, 10, 24, 27, 28, 43, 56, 84, 85, 96
white, men, 29, 41, 48, 50, 53, 54, 78, 79
white, women, 11, 24, 31, 42, 46, 49, 51, 53, 56, 80, 150, 173, 184, 193
white flight, 3, 62, 179
white poor, 4, 26, 104, 105, 114, 116, 128, 134, 135, 136, 137, 138, 139, 145, 154, 187
working-class students, 10, 21, 24, 27, 28, 38, 43, 46, 56, 84, 85, 96, 97, 133, 139, 147, 166, 167, 168, 169, 207
see also class (economic status), lower-class

yearbooks, 20, 40, 41, 46, 47, 61, 85, 97, 120, 121, 123, 126, 138, 157, 161, 171, 172, 173, 180, 181, 184, 188, 189, 203, 212

zero tolerance policies, 8, 12, 99, 113, 115, 127, 137, 138, 141, 145, 149, 181

GPSR Compliance
The European Union's (EU) General Product Safety Regulation (GPSR) is a set of rules that requires consumer products to be safe and our obligations to ensure this.

If you have any concerns about our products, you can contact us on

ProductSafety@springernature.com

In case Publisher is established outside the EU, the EU authorized representative is:

Springer Nature Customer Service Center GmbH
Europaplatz 3
69115 Heidelberg, Germany

www.ingramcontent.com/pod-product-compliance
Lightning Source LLC
LaVergne TN
LVHW011814060526
838200LV00053B/3779

Eddie nodded, once with hesitation, then more vigorously.

Jeff was right. Let's do this!

He turned the car into an estate with generously sized houses and expensive cars, and pulled up outside a large family home. Red brick. At least five bedrooms. White picket fence. Similar to where he'd grown up in New Jersey.

Jeff's hand was on his gun and he was out of the car before Eddie was even able to take the key out of the ignition.

"Come on, kid," he said, and Eddie followed, drawing his gun, refusing to feel any reluctance.

They strode down the garden path, reached the door and stood either side. It was slightly ajar. Jeff knocked on it. No answer.

He gave Eddie the nod.

"Police! Anyone there, make yourself known!" Eddie shouted.

They waited.

Silence.

Jeff gave Eddie another nod and he pushed the door open. It creaked into a house lit only by the few shafts of light that managed to seep through the cracks of the closed curtains.

Eddie stepped in, pointing his gun to the right side of the hallway, to the left, then up the stairs.

"Police, is anyone here?" he shouted.

Nothing.

"You take left, I'll take right," Jeff said.

"Roger."

Eddie turned to his left, passing a few family photos on the walls, and entered the living room. A large room with an expensive suite and a fireplace.

He almost choked on the sight.

Four bodies. A woman. Two girls. And a boy. Coated in fresh blood that glistened in the light. Their eyes staring at the ceiling. Their wounds gaping open like their skin was screaming.

"Shit..."

He edged into the room, checking all the corners, all the shadows. As he kept his gun ready in his right hand, he reached down with his left and checked the neck of each body.

No pulses. But the bodies were still warm. They hadn't been killed that long ago.

He edged forward, through the living room, and into the kitchen.

Another body lay across the entrance. This time it was an old lady.

So that's his wife, three kids, and his grandma.

Who else was left?

He walked through the kitchen and into the dining room.

The man was sat so still at the table that he didn't notice him at first, and when he did, he jumped.

"Freeze! Don't move!"

But Trevor Paul Scott wasn't moving at all. He was motionless, nursing a cup of tea, his face covered in his family's blood.

He smiled. Gave Eddie a nod. Sipped his tea.

Jeff entered from the opposite side.

"Got five deceased in here," Eddie told him.

"Got three in here," Jeff responded.

Sirens grew louder in the distance. Backup was close.

"Let's get him cuffed and ready," Jeff said. "Go ahead."

Trevor Paul Scott presented his fists, ready to be handcuffed.

Eddie glanced at Jeff's gun and checked it was aimed on

the suspect before placing his gun on his belt and approaching. He placed one handcuff on the suspect's wrist, and then–

The suspect moved so quickly Eddie was on the floor, grabbing his throat, before he'd realised what had happened.

His gun had been taken. From his belt. Fired at Jeff. Jeff was on the floor. The suspect had something. In his hand. A blade.

It was the last thing Eddie saw before the man slit his throat and he suffocated at his feet.

By the time more officers arrived and successfully arrested the suspect, Eddie's body was limp.

AFTER

CHAPTER TWO

She lifted her finger and traced the outline of the clouds. Strange things, clouds – they look like everything and nothing all at once. They are both existential and non-existential. Distinct yet formless.

Cia remembered walking past a couple in the park when she was young and hearing them discussing the clouds; what shapes they could find, how they might have been formed, romanticising what they were. Her dad, always a scientist, simply explained that they were a mass of water droplets suspended in the sky.

She remembered thinking that her dad's explanation was far more beautiful. Why try to create magic in something that is already so very magical?

That's when Cia realised she was awake. And, with this realisation, came the knowledge that she had been asleep.

And she was moving. Yet she wasn't. She was lying on something hard and wooden, completely still. Yet that thing was swaying. Drifting gracefully along a current, pulling her gently back and forth.

It was a boat.

She lifted her head.

It was a motorboat. She could see the handle that controlled the direction of the propellors.

Boy was sat by it, though he wasn't controlling it. His head was in his hands, and his body was furiously shaking. He was muttering something. Something like, "No, no, no, no, no…"

"Boy?" she said.

He gasped and lifted his head. His eyes met hers and he wrapped his arms around her so keenly that it made the boat wobble.

She patted his hair and, in the strength of his grasp, realised just how much she ached. Her arms were heavy and weak, stabbing pains gripped her chest, and her head was pounding. A scalding pain down her belly was accompanied by a line of blood oozing through a rip in her t-shirt. She wanted to look at it, but also knew that she was probably concussed as well as injured, and should avoid sitting up too soon.

"Okay," she said, enjoying his affection but not enjoying the pain. "It's okay."

He didn't move. He wouldn't take the hint. He wasn't the kind of boy that could read social cues – not that he needed to. His autism was irrelevant in a post-apocalyptic world full of monsters. So long as he knew how to survive, and did as she instructed, he could achieve all this world needed him to achieve; to stay alive.

"Thanks, but you're hurting me."

He still didn't get the hint.

"Boy, I need you to get off me."

He loosened his grip. She smiled at him and pushed herself up, but only a little, just enough so she could rest on her elbows and see where they were.

"Shit…"

She looked around and peered into the distance, but there was nothing. Just water. Lots of it. An ocean surrounding her. No sign of the coast.

How did she get here?

She rubbed her head, then rubbed her eyes. Willed the headache away. Lifted her t-shirt slightly and looked at the open wound across her belly. It wasn't too deep, but blood was still seeping from it.

Then she looked at Boy, who was still staring at her, expectant, like she was going to have all the answers.

It was the same way he always looked at her.

"Where are we?" she asked.

He shrugged. "The sea," he answered.

"How long have we been here?"

She realised as soon as she asked the question that Boy would not know. Cia always figured out the time of day by the position of the sun in the sky and, looking upwards, she could see the sun sinking over the horizon. Very soon, that sun would set, and they would be shrouded in darkness, with no idea what was around them.

At least no Thoral or Lisker or Waster or Maskete could get them here.

Then again, that wasn't strictly true. A maskete could fly by, swoop down, take one of them in its claws and deliver them to its young for dinner.

And there were the monsters that existed long before Hell opened and unleashed these beasts. Sharks, for example. If the boat was in the middle of the ocean, then they were probably quite near sharks.

How much petrol was left in the boat? Would there be enough to get them to the coast? And which way was the coast?

"Boy," she said, slowly and particularly. "This is very important, and it is not to have a go at you, I just need to know – which way did we come from?"

He looked confused.

"I mean, which direction did you take us? Which way?"

He looked blank.

"Which way, Boy? I need to know."

His eyes welled up. He turned away, burying his face between his knees, surrounding his head with his arms.

He didn't know.

He probably hadn't even taken them here. He'd probably just let them drift.

Which forced Cia to return to the question... how on earth did they get here?

She tried to think. Tried to remember. Tried to recall what exactly had happened before she woke up in this boat.

Right. Come on. Let's start at the beginning.

What had she done when she woke up that morning?

They had been in a city. On the south coast. Brighton, perhaps. Yes, it was Brighton. She remembered the pier – or, at least, what was left of it. Most of it was a charred, burnt-out wreckage on the beach below.

They'd hid in one of the buildings. Near the Sea Life Centre – she remembered because Dad took her there when she was a kid. Whilst other children were marvelling at the sight of the fish, she would be enthralled by Dad's commentary, telling her all about each animal, the science of how they lived, and how they survived. About how they breathed with their gills. About how some lobsters could live up to twenty years. About how catfish have 27,000 taste buds, while we have only 9,000. Then he'd ask why – "Why would evolution give them those taste buds? What would they need them for?"

Then he'd leave the question hanging, and she would wonder, and wonder, and wonder.

That morning, they had walked past the Sea Life Centre; or, at least, what remained of it. But why had they walked past it?

They were looking for food.

And then they'd heard a growl.

Cia's body stiffened.

She remembered.

CHAPTER THREE

Cia's head turned.

Did she hear that?

A growl.

Yes, she did. She was sure of it.

She turned her attention to Boy, whose eyes looked widely back at hers. He'd heard it too; the look on his face made that clear.

She looked around for cover. They were beside the beach, a mass of pebbles leading to the sea. They could run further into the city centre, where they could be more concealed. Maybe even find a building to hide in.

Another growl. Louder. From the city centre. The exact place she was looking to hide.

She looked at Boy. Tried to think.

This was ridiculous.

They had survived this world for a long time now. She'd fought monsters numerous times, and had to do unspeakable things to other humans to thwart humankind's demented nature – so why did the growl of a Thoral still startle her?

The road shook. The grand smack of a giant paw on the ground reverberated through their bodies.

It was close.

The buildings trembled. The pounding of its steps grew closer. They didn't have time to run into the city. It was going to emerge from the buildings at any moment, and they were exposed, and after everything they'd gone through, they were going to be killed and eaten by a damn Thoral just strolling around Brighton.

No. No we are not.

She grabbed Boy's wrist and dragged him away from the buildings, toward the beach, away from the continuous tremors of the Thoral's steps.

She looked around. What about the rubble of the pier?

No, it was too far away. She could see it in the distance, but there was no guarantee they could get to it in time.

There was a slight drop from the path to the pebbled beach below. Maybe they could use that for cover?

No, it wouldn't work. The Thoral would smell them if it didn't see them.

Its shadow grew closer.

She had no more time to think. She clutched Boy's wrist and ran toward the sea.

How did Thorals feel about water? If they didn't like it, could Cia and Boy get far enough into the sea that they were out of reach?

But what about Boy? He was young when she'd found him, probably too young to have finished swimming lessons. Even if he had, there was no guarantee they had worked.

"Can you swim?" she asked him.

He didn't say anything, but a slow whine and the shake of his lip was enough of an answer.

She looked around again, frantically, searching.

Screw it, she thought. She ran, dragging him with her across the pebbled beach.

The Thoral leapt out from between the buildings, forcing the ground to shake with such ferocity it forced Cia and Boy to their knees.

In its landing, the Thoral barged into a nearby building, and bricks tumbled down its walls.

Cia made the mistake of looking over her shoulder.

It had seen them.

With a few measly steps it would be behind them.

Cia grabbed Boy, pulling him, sprinting hard, her heart punching her chest, her lungs struggling against the exertion.

She saw something in the sea. Their target. Far out from the coast, but close enough that they could wade through the water and the sea's surface wouldn't go above Boy's neck.

A motorboat.

"The boat," she told Boy. "Get to it."

With their eyes locked on the boat, they ran, reaching the water's edge.

The Thoral's claws landed on the ground behind them, and the rumble sent them once again to their knees.

The Thoral lifted its front claw, setting its eyes on Boy.

Cia grabbed Boy, pulled him out of the Thoral's reach, and put herself between the tip of the claw and Boy.

The tip of the claw swiped across Cia's belly, leaving a trail of blood and a rip in her t-shirt.

She fell to the ground, momentarily unable to breathe, or to even think. The pain was excruciating, but she manage to push herself up and wade through the water.

The claw lifted once more, hit her head, and sent her onto her back.

Her eyes closed, then opened, then closed and opened.

She was awake, but she was not aware.

She was clinging desperately onto her consciousness, knowing she was about to black out.

CHAPTER FOUR

Boy grabbed Cia's arm and tugged.

What was she doing?

Her head rolled to the side. A moan came out of her, a groggy whine.

Meanwhile, the Thoral was preparing for another swipe.

Boy wanted to crawl into a ball and bury his head. That was what he normally did; close his eyes and hope that it will go away. Normally, by the time his eyes opened, Cia would have gotten rid of whatever was being scary.

But Cia's eyes were closing and he had to be brave.

This was Cia – or Rosy, as he'd come to call her – and she always persevered.

He forced her to her feet and, groggily, she waded into the water with him. She wasn't balancing well, and she kept toppling over, but he managed to keep her steady.

The Thoral went to swipe again, at which point Cia looked over her shoulder, saw it, grabbed Boy, and pulled him under the water.

He didn't like being under the water. He'd never been

under water before. He couldn't breathe and everything felt heavy.

He lifted his head back above the surface. The Thoral remained at the water's edge, not coming any closer.

It won't go into the water...

They had to get far enough into the sea to be out of its reach. Boy then realised that Cia hadn't come up out of the water.

He reached around for her, finding her head, and pulling her up by her t-shirt. Her eyes were closing.

"Rosy! Rosy, stop it!"

Her eyelids parted, only slightly, enough that he could see her pupils, then closed again.

He put his arm around her and dragged her body further into the water. Her petite frame was fairly easy to carry with his large, lanky physique; for a little bit, at least.

The Thoral swiped again.

Boy really, really didn't want to go under the water again, but, in the split second he had to make the decision, he knew he had no choice, and he pulled Cia under with him.

He waited for a moment, then lifted his head above the surface, pulled Cia up with him, and dragged her onwards.

The Thoral swiped again, but this time it couldn't reach, and it would not enter the sea. Boy could see the anger in its eyes, like it was cross at their impudence. How dare they get away; how dare they find a way to leave its reach...

So, it waited. Watching them. Anticipating their inevitable return once they found that the ocean gave them nowhere to go.

But the motorboat wasn't far.

Cia hung off Boy. He held her arm in place around his shoulder, but her body was otherwise dropping. The longer he dragged her, the heavier she became. He couldn't keep carrying her; not whilst also wading through water.

He let her go, lifted her feet until she was floating, and supported her body as he guided her toward the boat.

Boy hated this. The Thoral was growling. The water was making Boy panic. His anxiety was attacking him on all fronts and he just wanted to break down and wish it all away.

But Rosy needed him.

He wiped water from his eyes, unsure whether it was sea or tears. He put his arms beneath her torso and, with a huge push, shoved her over the side of the boat, then grabbed her legs and shoved them over too.

He followed her in, careful not to capsize the boat, then collapsed, relieved to be out of the water.

Rosy lay on the wooden base of the boat, her eyes closed. Was she dead?

Boy was sure she wasn't. Her chest was still moving. He placed his hand over her mouth he could feel her breath against it.

The Thoral roared again. It was fuming, and it wasn't done yet. It lifted both its claws into the air and, in one devastating blow, swung them downwards, into the ground, which caused a tremendous wave that pushed them further out to sea.

It made Boy feel like he was about to be sick. He lay down next to Rosy, put his arms around her, and held on with everything he had as the boat bounced, one way then the other, soaring across the surface, away from the coast.

Boy just closed his eyes and hoped it did not capsize.

The wave carried them a long way, and it kept carrying them, until the water was too deep for him to wade through; until Brighton and its beach were in the distance.

Eventually, the wave calmed, and they were just drifting across the ocean.

The sun was halfway down the sky at this point. By the time Cia awoke, it had almost set.

CHAPTER FIVE

Cia cupped Boy's face, looking him in the eyes.

"You did well," she told him. "So, so well. Really, you did."

He nodded.

"I mean it. You were so brave."

He nodded again. She could see the distress in his face. No praise would undo the trauma of going under the water, escaping the Thoral, getting her into the boat, and drifting along the current with no idea when his protector would wake up.

She put her arms around Boy and held him close, knowing that was what he needed – her touch would comfort him, not her words. She felt him sob into her shoulder, but she did not release him; once she released him, she was going to have to figure out what to do next.

Night-time was imminent, and there would be nothing to illuminate their predicament. No torch, no light, no candle – just the vague light of the moon.

Could they wait in darkness for the entire night?

Boy probably wouldn't be able to take it – and she wasn't sure she could, either.

She released him. Smiled at him. Ran a hand down his hair.

"It's going to be okay," she told him, knowing she had no way of following through on such a promise.

She could drive the motorboat. Go in one direction and hope it's the right one. Surely, if she just drove forward, they would reach somewhere. But what if they ran out of petrol before they did? Then they would be stranded, even further into the ocean.

If she was going to steer the motorboat, she'd need to know where to steer it to.

Right, come on. Figure this out.

Dad had taught her all about science, but not about geography. Right now, it felt like it would have been a good lesson for her to learn.

She knew the sun rose in the east and set in the west. Therefore, if she directed the boat toward the setting sun, they would be heading west. So, if she drove to the right of it, they would head north.

Wouldn't that take them back to England?

But she had no idea which way they had drifted already. They could have drifted far into the west, which meant that if she went north, they would end up stranded in the North Atlantic Ocean.

She rubbed her sinus. Her headache was getting worse.

She took Boy's hand, knowing he was staring at her, but holding it as much for her own comfort as his.

"We'll figure it out," she reassured him, but her smile wasn't genuine, and she wondered if he was picking up on that. He struggled to read body language, but he wasn't an idiot – surely he could tell she was lying.

Then she saw it. Far in the distance, a small speck slightly lit by the sinking sun.

Something.

Was it land?

No. It was too small.

Was it a ship?

It was. It was a ship. It had to be.

She felt relief. Their salvation was here. They were saved! They just had to get to the ship and hopefully there would be people on it and–

Then she realised – it wasn't that simple, was it? All the other survivors they'd come across were conniving, or psychotic, or both.

More recently, she'd had to save Boy from a group of people wanting to sacrifice him to the creatures. Before that, she'd almost been killed by a man she thought she loved. Before that, her father had left her for dead.

Boarding that ship could be even more dangerous than hanging around in the middle of the ocean.

What if they were pirates, still sailing around from before the creatures arrived? Pirates nowadays didn't have peg legs and swords – they had machine guns.

Or what if they were another cult? Another group of people who'd use her for their sexual gratification, or use Boy as a slave?

Then again, what if they were good, honest people? They could be families who'd fled their country to protect themselves from the creatures.

She shook her head. Such people didn't exist. At least, she hadn't found any.

She glanced at Boy. He had his arms wrapped around his body. He was shivering. His breath was visible in the air.

She realised how cold she was as well, having been too

preoccupied with her thoughts to notice. The wound on her belly was hurting more intensely, throbbing under the added pain of sub-zero temperatures.

They had no choice. If they stayed here overnight, they would freeze to death.

She pulled the string on the tank, bringing the propellor to life, and guided them toward the ship, without a weapon or hope, and with no idea what she was about to find.

CHAPTER SIX

THE CLOSER THEY came to the ship, the more apparent it became just how large it was. The hull was a dark grey, whilst everything above it was a lighter whitish grey. Windows appeared all along the side, home to many rooms, but also many places to hide. It appeared to be a cruise ship, and it was still moving, but only slightly.

Cia was beyond cautious; she was pessimistic. She had no doubt that, with so many places to hide, there could be many opportunities for someone to kill them before they could fight back. Or worse – she'd once come across a group of men forcing their women to repopulate the earth; there were no depths to the depravity that humanity could sink to.

But, as much as she hated it, she had to remind herself that they had no choice. If they waited until morning, they were likely to die from frostbite, or sharks, or an unanticipated storm that would throw them into the water.

Boy couldn't swim. If he went overboard, he'd drown. This was their only option.

At the frontend of the ship, she found a ladder. But she didn't climb up yet.

Instead, she waited.

And she listened.

And she heard... nothing.

No talking and no movement. It was night, so maybe everyone was asleep – but surely there would still be people steering the ship, or keeping watch, or something. The ship was moving, and the engine was rumbling – although it was moving very, very slowly; almost like it had been left that way.

"Hello?" she shouted, against her better judgement.

There was no answer; just the sounds of the waves.

She checked her belt for her knife. It wasn't there. She must have lost it in the struggle, or when Boy was pushing her into the boat.

That meant she was unarmed.

She looked at Boy, considering whether to climb aboard, and noticed he was shivering. Furiously rubbing his arms, trying to warm up. He was too cold; she couldn't expect him to stay here any longer.

"Come on," she said, reaching for the steps. She took hold of them, only to find her hand slide off. There was something sticky on the steps. Whether it was blood, or some kind of gunk, she wasn't sure. She was going to need to let Boy go first in case he slipped; she might not be able to catch him, but she could certainly break his fall.

"It's slippy," she told Boy. "So you're going to need to grip. You go first."

Boy reluctantly took hold of the ladder, his body shaking, and Cia supported him as he climbed up.

She followed behind, firing through worst case scenarios of what might happen to them once they'd climbed aboard; but, as they climbed over the railings and fell onto the deck, no one appeared.

Cia glanced over the railings and watched their motorboat drift away, disappearing into the ocean.

They were stuck here now.

She waited for someone to come running out, perhaps armed, challenging them, demanding to know who they were and what they wanted.

There was nothing. Just the quiet whistle of the wind and the flapping of a flag at the top of a pole.

Cia didn't like it. Not one bit.

"Stay behind me," she said, and edged forward.

They approached a cabin that contained the large steering wheel – or was it a helm? She had no idea how she knew that, but she did.

As she grew closer, she wondered – what was that on the window?

But she knew what it was.

It was blood.

It didn't glisten in the moonlight. It wasn't new. It was crusted, fixed to the glass.

She stopped by the door. Placed her ear against it. Listened.

Could there be something in there?

But what? Thorals and Liskers and Masketes would be too big and noisy, and she would certainly hear a Waster.

Could there be a person? Someone who'd seen them and was waiting?

"Wait here," she told Boy, let go of him, and placed her hand on the handle.

She readied herself for a fight, prepared herself to storm in and dodge whatever came after her.

But when she opened the door, no one came.

All she found was a dismembered man, his eyes wide

open and his chest split apart, as well as his legs and his arms and his throat... In fact, it appeared that each limb had been shredded.

This person had died in a horrific way – and it did not look like a human had done this.

CHAPTER SEVEN

Cia left the cabin quickly, taking hold of Boy's arm and pulling him along the deck, looking from one window to another. Everywhere she went, she found the same thing.

Bodies. And lots of them.

Inside the cafeteria. Corpses with torn-apart muscles draped over tables and chairs. Bedrooms with bodies piled on top one another. Corridors marked with blood. Loose limbs that no longer belonged to anyone.

Each body had been opened and fed upon, until all that was left was shredded muscle and bone.

What could have done this?

She needed to know who these people were, and why they had been on this ship, and how long they'd been out here. This would help her determine the possibility of whatever did this coming back.

She took to the stairs, descending a few floors and entering another corridor, looking for clues. The bedrooms and offices disappeared, and all she could see were rows of doors with small, barred windows.

They were prison cells.

Mostly empty. Covered in streaks of red. Some bodies had been devoured just like the others. Some had evidently starved to death, unable to free themselves from their cell. The entire corridor smelt like rotting cabbages and shit.

Boy began to whine. He was close to having one of his breakdowns, she knew it; the remnants of prisoners inside these cells were horrific, and he didn't understand his own fear well enough to know it was upsetting him.

So she kept them moving.

Eventually, they came to the end of the corridor and found some offices. Each had a name fixed to the door, each with a title beneath them, such as *Jane Knowles, Psychology* and *Lee Harvey, Floor E Prison Officer*.

Then she came to what she was looking for. An office that belonged to *Jason Arms, Captain*. Inside was mostly untouched, with only a few specks of blood and no body. She pulled Boy in, sat him on a seat, and knelt before him.

"Are you okay?" she asked.

He shook his head.

"Yeah, stupid question." She looked around, trying to think of something that could occupy his mind while she searched for information. There was nothing.

"Listen, I need you to wait here while I look around, okay?"

He started to panic, so she quickly added, "In this room. I'll be looking in this room, so I'll be right here. But I need you to be patient, okay?"

He paused, then nodded. She knew how much it took for him not to panic. She gave him a solemn smile and turned to the desk.

She sat in the large, leather office chair and opened the drawers. Inside were folders. She took them out, to find each one entitled *Prisoner Transportation*.

Was that what this ship was doing? Had they converted this cruise ship into a ship able to transport prisoners? But why and where to?

She opened the folders and looked through them. Inside was a spreadsheet of prisoners names and their crimes. Most were in for convictions like theft or drug dealing, but a few stuck out; a few that were in here for rape and murder.

In a way, she was relieved they were dead. This information didn't make the ship feel any safer.

She looked through the rest of the drawers, finding administrative documents, such as staff rotas and medical forms. Nothing useful.

She turned back to the desk and covered her face. What were they going to do? She could hardly fall asleep thinking they were safe aboard this ship, but they had nowhere else to go.

A coat lay over the corner of the desk, and she noticed a few pieces of scrap paper poking out from beneath it, so she pulled the coat away to find a note stuck to the table by a knife. She pulled the knife out of the desk, tucked it in her belt – it might be useful later – and looked at the note. The handwriting was scruffy, but she managed to decipher the words.

Should someone find this note, please pay attention to what I have to tell you.

You do not know everything.

They told you there were four monsters. But that was just at home, that was just what we thought we knew, that was

The last *s* of *was* turned into a long line, trailing across the page.

On the next scrap of paper she found a single sentence, scribbled recklessly, and repeated over and over:

There are more.

There are more.

There are more.

CHAPTER EIGHT

Cia needed to get out of that room, desperately, so she grabbed Boy and dragged him outside into the moist ocean air and, without so much as a word, collapsed to her knees.

There were more?

How could there be?

She had spent the last few terrifying, arduous years battling the four monsters she knew of, trying to just stay alive. She'd learnt about them, known how to evade them, to hide from them – and now what? There was another creature?

And, what's more, it was a creature that could do this to an entire ship? Leave it covered in bloodshed and death and mayhem like it was nothing?

Her stomach lurched. There was not much in it, so when sick reached her throat and forced itself out of her mouth, only blood and bile landed on the deck.

She felt Boy's hand on her back. He was terrified, she knew, she could feel his arm shaking, but he was being strong. For that, she was grateful. She really needed him to be strong.

She gathered herself. Pushed herself to her feet. Ran her hands over her face and through her hair. Gazed into the vast ocean, dimly lit by moonlight. The waves were gentle, though she knew they wouldn't always be. Whilst those waves were barely rocking the ship now, a storm would eventually come – and that storm would be torrential.

And this monster... this other creature... could it travel through water?

Of course it could. *Don't be silly*. How else would it have done this to a ship load of people?

Did that mean the monster could be out there right now, beneath the surface, hunting them? Could it smell them? How were they supposed to escape it in the middle of the ocean should it attack them?

"We can't stay here," she announced, without any conscious decision to speak.

But then came the next thought – how would they get back to land? Where exactly was land?

She rubbed her eyes. She was tired. It was dark. And the wound across her belly the Thoral had left was throbbing. They could do nothing until morning, so the most productive thing she could do at that moment was get some sleep. It was a risk; staying on this boat, shutting her eyes – but it would make no difference. If a creature that destructive came along, there would be nothing she could do whether she was awake or not.

But what if there were prisoners still alive on the ship?

She made a decision. They would find a room. Somewhere with a bed, somewhere less blood than the other rooms she'd found. They would barricade themselves into that room so nothing could try to get in without making a raucous amount of noise. As soon as it was light, they would wake up, and they would try to find out where they were.

Maybe she could steer the ship – it would be safer than the motorboat.

"Come with me," she told Boy, and led him back into the corridor.

They needed to get deeper into the ship, as far into its centre as they could, and make it harder for something to smell them. It probably wouldn't make much difference, but she was going to do anything that could improve their chances of survival, however small.

She led Boy through the corridor, to a set of stairs, and they made their way down. The next corridor contained cells, as did the floor below that, and the next one below that.

It was odd for a cruise ship to transport prisoners. Where were they transporting them from and to? Why would prisoners need to go to another country?

The most likely reason was that they were evacuating. Perhaps this ship had left when the disaster struck. The government would have had a responsibility to evacuate and protect anyone in their care, and that included prisoners. It could even be from a prison in another country, which led her to wonder – how much were other countries affected? Did the monsters hit the UK after other countries, before, or at the same time? And how well did they deal with it?

They were questions she'd never entertained before, which made her feel a little stupid – why hadn't she thought about other countries and the state they were in? They may have had more success fighting the creatures. Who knows?

No. She was getting carried away. Chances were these creatures hit the whole world and everyone suffered.

Then again, didn't the note imply that different countries had different monsters? What if the creatures were indigenous? What if another country's monsters were easier to deal with?

She shook her head. She had a headache and was too tired for all this thinking. She descended another flight of stairs, only to find more prison cells, so continued further down. She wanted to be safe, but she didn't want to sleep in a cell. Not that she'd care, but she wasn't sure how Boy would take it.

Yet the next floor down only produced more prison cells. She punched the wall in frustration, and regretted it the instant her knuckles collided with metal. It caused a clang that echoed down the corridor.

She leant against the wall. Wished they were somewhere else. Anywhere else.

And that's when she heard it.

Faint, at first. But when she listened, really listened, she knew she could hear it.

In the distance.

Somewhere along the corridor of prison cells.

A voice, shouting, "Hello? Is anybody there?"

CHAPTER NINE

With one hand gripping her knife and the other clutching onto Boy, Cia walked towards the voice.

With each step, it grew a little louder. It was definitely somewhere along this corridor.

"Hello? Is there someone there?"

"Stay behind me," she told Boy, and edged forward.

She passed prison cells on either side, most of them open; solid white doors with small windows, and more bodies left ripped apart.

She felt Boy's head turning, his eyes wandering, and she gripped his wrist to draw his attention.

"Don't look in the cells," she said. "Just look straight ahead."

She had no idea if Boy was following her instructions, as her wide eyes were fixed ahead, ready for whatever was to come.

The voice grew louder.

"Hello? Please, is someone there?"

The corridor curved and the voice became clear.

"Hello?" it continued.

Cia paused, waited, then reluctantly answered, "Hello?"

"Oh my God, who's there?" The voice became excited. "Oh thank Christ, someone's there! Where are you? Please, help me!"

The voice was coming from one of the cells.

Cia stopped. Waited. Wary.

If all of these other prisoners had been killed, how was this one still alive?

He could be feigning imprisonment. He could be luring them into a trap. For all she knew, he could be a murderer.

"Hello?" he said. "Are you still there?"

"Who are you?" Cia shouted, keeping her voice assertive and strong.

"My name is Paul," he said. "Please, help me."

He had an accent. One that wasn't British.

"How are you still alive, Paul?"

"Wh – what?"

"You heard the question."

A moment's silence passed.

"I don't understand, I... I don't know how to answer that question."

"Well, Paul, every other cell contains dead bodies. Why are you the exception?"

"They are dead? All of them?"

"Answer the question."

"I – I don't know. I don't know how to answer that."

"How have you managed to eat?"

"I..." his voice trailed off. "Please, just help me. I'm so hungry. I'm so thirsty."

Cia sighed. She stared at the cell the voice was coming from.

"Do you have any weapons?" she asked.

"What? No."

"Are you able to get out of your cell?"

"No!"

"Why not? Most cell doors are open, why isn't yours?"

"I don't know! I guess everyone was killed before they got to mine!"

Cia felt unsure. She didn't trust him.

"Stand back against the wall, away from the door. If you try anything, I will kill you."

"Yes. Okay."

A few steps came from the cell.

Cia looked over her shoulder at Boy. His face was painted with terror, his lip quivering. Strange, really, how she'd become so accustomed to this world, yet he'd remained so vulnerable to it.

Then again, that was one of the things she loved about him the most.

"Wait here," she told him.

"No!" he objected.

"You'll still be able to see me. I'll just be over there."

He kept hold of her arm until she was out of his grasp. She edged toward the cell slowly, her knife at neck height, ready to strike at the first sign of danger.

When she reached the cell and peered through the cell bars, she saw a man standing against the far wall, just as she'd instructed. His skin was covered in tattoos; beside his right eye were two tear drops, and beside the other eye, an umbrella. She could also make out some large tribal tattoos coming out from beneath his prison uniform.

She tried the cell door to see if it would open. It wouldn't budge.

"What have you been drinking?" she asked.

"What?"

"Stop playing stupid and answer my question."

"I... My..." He dropped his head. "My piss."

He didn't meet her eyes.

"And what have you eaten?"

He still didn't meet her eyes.

"Paul, I'm not going to help you unless you're honest with me."

"I – I did what I had to, all right?"

"What does that mean?"

"I did what anyone would to survive. I didn't want to, but there was no choice, it was the only way..."

"What did you do?"

He kept his head bowed, closed his eyes, then nudged his head in the direction of the other side of the cell.

She shifted to the side and covered her mouth as she gasped.

On the floor was, what appeared to be, this man's cell mate.

Or, at least, what was left of him.

Much of his skin and insides were gone, leaving his bones naked and exposed.

Cia looked back to Paul, who still couldn't meet her gaze, as she realised what he'd done.

"Oh my God..." she whispered.

CHAPTER TEN

Cia shuddered and suppressed the need to be sick.

"I didn't have a choice, okay!" Paul insisted, becoming more passionate.

Cia turned away. She couldn't look. Despite everything she'd seen, she couldn't look.

"He agreed to it!" Paul said. "He said it had to be one of us, and we decided who would survive and who would... you know..."

"How?"

"What?"

"How did you decide?"

Paul took a moment, clearly deciding whether to tell the truth.

"Rock, paper, scissors," he finally answered.

"You did what? You decided who would live and die with a game of rock, paper, scissors?"

"We did exactly what you would have done!"

Cia frowned and shook her head in disgust.

"I've survived, haven't I?" he said.

"There are some things that aren't worth survival."

"You think I liked it? You think I like drinking my piss and eating my friend? I did what I had to do, for fuck's sake!"

Cia looked at Boy, who was shivering, staring at Cia. He really didn't want to be here; she could see that. Neither did she, but at least she could recognise the depravity of the situation. Whilst Boy may understand it, it was his inability to recognise what he understood that caused him so much anguish.

"Please let me out," Paul said, his voice quieter.

"Are you kidding me?"

"Please. I'm starving. I've been in here for so long."

"And what about us, hey? What about when we run out of food? Are you going to paper, rock, scissors with us?"

"Please, I promise you I would not do that."

Paul stepped forward, and his features became clearer. He looked both young and old at the same time. His body slumped with both feebleness and malice. He stunk, a combination of piss and shit and body odour. She knew she didn't particularly smell of roses, but this man's odour was foul.

"What are you in here for?" she asked.

"What?"

"I said, what are you in here for?"

"I don't know what you mean."

"This is a prisoner transport. What are you in prison for?"

The man went to speak, then shook his head, apparently unable to tell her.

"You have two tear drops tattooed beside your eye," she said. "What do they mean?"

"Please, they are just tattoos I got done in prison. The more you let other people tattoo you, the less shit you get."

Cia watched him. He looked desperate and, if this were a

few years ago, when she was new to this post-apocalyptic world, she probably would have found a key and let him out.

But she wasn't a child anymore.

She had witnessed the depths of what the end of the world compelled people to do, and whilst this man may in fact end up being harmless, and may well have just done what he had to do – Cia was not prepared to take the risk.

"I'm sorry," she said. "I can't help you."

"No, please, I won't hurt you, I'm harmless, I promise."

He sounded so genuine that Cia was almost tempted to believe him. Then she looked again at what remained of his cell mate, and she was resolved.

She turned without a second thought and marched back down the corridor, Boy instinctively following.

Paul's voice shouted again, persisted in cajoling her, and she ignored it.

That is, until she heard him say, "I can help get you back to shore! I know how!"

She paused, just for a moment, tempted to go back.

Did he really know how to get them back to shore? Did he know which direction it was?

Was it worth the risk of releasing him? Getting back to the shore would mean nothing if he killed them…

She shook her head. Other survivors had always proven to be the biggest threat.

She walked on.

She and Boy found a cabin with beds. They collected desks and chairs from other rooms, which they used to form a barricade against the door. Once they were done, they lay on the bunk beds – Cia on the top, just in case Boy was to toss and turn in his sleep and fall off – and closed their eyes.

Despite the horrors of the day, Cia fell asleep, her fatigue too much for her anxiety to fight against.

CHAPTER ELEVEN

THE CORRIDOR WAS dark yet full of flashing lights, though she couldn't see a light anywhere. Specks of water landed on Cia's face, like a storm had found its way inside, which was impossible – but Cia accepted it.

She found an open cell.

She walked inside.

Saw a man. A young man. She recognised him. She'd killed him with an ice pick. She'd fucked him because she had to, and as he came, she'd stuck it into his chest.

This had happened a long time ago now. Back when she'd lost Boy and was trying desperately to find him. Back before she'd unleashed monsters on the sanctity.

Without any awareness or control over what she was doing, she stepped forward. He lay on the bed, naked. She mounted him, finding herself naked too, and let him enter her.

It had been her first time, and it hurt just as much now as it did then.

She lifted the ice pick again.

Stuck it into his chest.

Except, this time, he didn't die. He looked up at her with a patronising smile, tilted his head, and said, "Cia, why are you like this?"

To her side, someone walked in. Daniel Rose.

She looked back to the man she was riding and he was gone. She was sitting on a chair in an office. Her father's office.

The father she had left to die.

He sat opposite her. It felt like a therapy session. He smiled like he was about to give her bad news.

"Cia, I think you should answer his question."

"What?"

"His question. Why *are* you like this?"

"Like what?"

"... A monster."

"I'm not a monster. I've fought monsters!"

He chuckled. "Even those monsters wouldn't leave a desperate man to die in his prison cell."

He stood and left. She went to go after him, only for a strong hand to grab her arm.

"You left me to die, too, remember?"

She spun around.

"Dalton?" she said.

"Meh. What's left of me."

He stepped out of the shadow. His face was a bloody mess, half-missing, his skull exposed.

She gasped and turned away.

He laughed.

"You did this, you little wretch! You did this, so how do you have the moral high ground not to look?"

She shook her head furiously.

"Look at me."

She covered her eyes.

"Look at me!"

Against her own will, she lifted her head and looked.

Only now, it was Paul.

And his body was half-eaten.

"I'm sorry," she said. "I'm so sorry..."

Before she could say anything more, a large clatter ended her nightmare and she fell out of the top bunk.

The shock woke her. She lay on the floor, disorientated, looking around. Her back, which she had landed on, ached, a throbbing sensation shooting up and down her spine.

The boat had tilted to the side and, as it re-balanced, another tilt sent her sliding across the floor.

Boy was clinging to the pole of the bed.

"What's going on?" she asked.

Before Boy could answer, something pounded against the ship again, tilting it to the side, and, for a brief moment, Cia expected the ship to capsize.

Then it rebalanced.

Something was out there.

Something large enough to–

She broke from the thought as it pounded again, hard, into the ship, and the ship tilted even further and she clung onto the bed, holding onto Boy's arm as he fell out from under the covers. All the desks they'd used to barricade themselves inside almost toppled over them.

Once again, Cia expected the ship to capsize and immediately began trying to figure out what to do. They'd have to get out before the room flooded, would have to leave the ship while it sunk – but the ship re-balanced, shaking from side to side before steadying.

Cia rushed to the small, acrylic window and tried to see what had done this.

She saw a tentacle, then another, and something huge soared away into the distance.

It was unlike any creature she'd seen before.

They weren't alone. Just as the captain had written on those notes, there was something else out there. Another monster in the sea, capable of slaughtering an entire ship.

They needed to get to shore, and they needed to do it quickly, but she had no idea how.

She had no choice but to place her trust in a man who remained locked in a prison cell below them.

BEFORE

CHAPTER TWELVE

INITIALLY, the screaming had been constant. Persistent shrieks and cries of fear or pain, battering against his ear drums, invading his mind with everyone's incessant agony.

It was the first time Paul was grateful for being locked in a cell – even if he was locked in with such an abominable cretin.

Honestly, in all of Paul's incarcerated years – of which there had been many – Paul had never been given such a dull cell mate. What's more, Paul was pretty sure he was a nonce. Shay, this guy's name was, and all he ever did was read history books and write poetry and play chess – on his own, as well. He would sit at the board, playing against no-one but himself. Sometimes, Paul would 'accidentally' nudge the board as he walked past, or deliberately touch the chess pieces after he'd used their toilet.

That was one thing he never thought he'd get used to – sharing an open toilet with another dude. Sitting there, making eye contact with a guy squeezing out a shit as he tells you about his wife's latest letter.

Of course, circumstance was different now. The state

was responsible for the welfare of its inmates, so when the creatures attacked, they had a responsibility to ensure their safety; meaning they were evacuated while most people at home were being murdered and maimed. In a way, being imprisoned had given him more chance of survival than most of the other suckers.

Now, however, the cries of anguish and desperate prayers of all of those onboard made it clear that the ship was under attack. It swayed frequently, knocked to and fro, and on many occasions, Paul was sure it was about to collapse.

But it was a good ship. It hadn't betrayed them.

Then, as the night grew darker, the screaming stopped, and a far more menacing silence filled the corridors.

Either everyone had left, or everyone was dead.

He stood at the window to his cell, peering down the corridor. A few stains of blood marked the cream walls, but otherwise, there was nothing.

Whatever had attacked, had evidently finished, under the belief that it had killed or eaten everyone. Time went by, possibly a few days, but there was no way they could know; they had no access to outside light. It would be a long time before he saw the sun again.

Shay grew increasingly annoying, banging on about food and despairing as to what they were going to do. It didn't help. Paul was starving as well, and constant reminders that they were likely to die if they didn't find a way out of their locked cell weren't helping. Besides, he'd heard all the other inmates shouting to be let out of their cells, then once they were let out, he'd heard all the shouting as they were killed.

The cell was probably the safest place to be. If only he could figure out a way to find food without leaving it.

"What about drinking? We can't use the toilet water

forever, it's bound to stop working without someone to maintain it. We need to–"

"Would you *shut up!*"

Paul turned and kicked Shay's chess board across the cramped cell. The walls were closing in on him, and the small cell was shrinking, and he was panicking in the knowledge that he was not going to get out and would probably die in here and it was making him feel more and more claustrophobic.

"Honestly, set aside your ego," Shay said. "We have more important things to worry about."

"You fucking *what?*"

What had the little shit just said?

Paul unknowingly let a growl escape his lips. How dare he? Paul had murdered men for less.

"I was just saying–"

Paul grabbed Shay by the neck and threw him across the room.

Shay tried to get away from Paul, but there was nowhere he could go. Paul's already deteriorating mind was sinking further into insanity. He was in a tiny cell with no light and nothing to do and he wasn't sure he could last another hour in this cell alone, never mind putting up with this guy as well.

"We need to eat," Shay said, his voice shaking. "We need to think."

"I can't think. There is nothing to eat, there is–"

He paused. Looked at Shay.

That may not be entirely correct.

"What?" Shay said.

It would be so easy. Just two thumbs pressed into the scrawny prick's neck.

He'd been trying to change. He'd told his therapist that

he didn't want to be a killer. He didn't want to hurt people anymore.

But this was a new world.

A world where only the ruthless survived.

"Please, stop staring at me."

Paul stepped forward, looming over Shay, and mounted him. He placed his hands around Shay's neck, and Shay did all he could to fight him off, pounding Paul's arms, thrashing his legs, struggling with all the might he had.

But Shay was a weak little man, and Paul was experienced in murder.

Within minutes, Paul's hunger problem was solved. He had something that was going to last him a while.

AFTER

CHAPTER THIRTEEN

Cia's survival in this horrific world wasn't down to her tenacity, her instincts, or her determination – though those were all crucial aspects of her personality – it was her resourcefulness. She could create shelter out of twigs, create a hiding place with leaves, and create a weapon out of stones.

She was going to need to use such skills if she was going to gain control of the prisoner.

She really didn't want to let him out. She'd experienced the true depravity of human nature from people who had appeared kind, but this man didn't even bother with that disguise. To look at him was to look at a trap without its camouflage. Everything about him screamed *danger*, and she did not wish to put hers or Boy's lives at risk.

But, in the long hours when night turned into morning, as she lay awake listening to Boy snore, not wishing to fall asleep for fear of having more nightmares, she'd scrutinised their options.

Unfortunately, she could only think of one.

She had no idea where in the ocean they were, how to figure such a thing out, or even how to guide the boat home if she did. There was a monster in the ocean with them – one who'd almost capsized the ship with a few nudges – and if it was aware that more people had boarded the ship, it would be back.

They could not wait another day, never mind another night. She had no choice – but that didn't mean she was going to be a fool about it.

She told Boy to wait in the room and went to leave.

"No!" he said, leaping from the bed and clutching her arm.

"I'm just going to look around the ship, I need to find a few things – you'll be safer here."

"Please don't go."

"I promise, Boy, I'll be back soon."

With a bit of cajoling he reluctantly backed off, and she left before he could object any further.

She searched the corridors and the rooms, opening doors and looking from one mutilated corpse to another, until she found what she was after – the body of a prison guard. On their belt, she found handcuffs.

Next, she searched the supply cupboards, ignoring the bodies as best she could – she was used to the sight by now – and found some rope. The next part, however, she would need to improvise – as she was in need of a large, metal rod. She found a stick that was used to reach high windows in the cafeteria. It had an indent running inside of it, which was perfect, as the rope fit nicely inside of it. She fixed the rope to the stick and created a noose at the end.

Finally, she needed the key to the prison cell. She found this in an office at the end of the corridor with a plaque on the door reading *Sean Hamill, Prison Warden*.

With her items ready, she stepped lightly past rows of

cells. The smell of rotting grew stronger as she reached Paul's cell, and she felt more and more queasy.

Her gut was screaming at her not to do this. She ignored it; she had no other options.

She found him on the floor, huddled in a ball. Shaking. So stuck in his trance that he didn't even notice her.

"Paul," she said. He jumped. Leapt to his feet. The cell door shook as he fell upon it, and Cia backed away.

"You came back!" he said, practically crying. His eyes were odd. They were always wide and staring. Even now, whilst he was jubilant, there was something mad behind them.

Cia said nothing else. She wanted him to wait. She wanted it to be clear that she was in control.

"Please, I promise I won't hurt you, no matter what. I'm so hungry, would never hurt you, never would..."

She watched him. His movements were erratic. His limbs each seemed to work out of sync with each other. Lying in a cell for as long as he had, with the dead cell mate that he'd eaten next to him, drinking his own urine, was bound to make a guy go crazy – but she had a feeling he'd already been unhinged, and this situation had only exacerbated it.

"Surely you saw the creature last night?" he said. "It will be back... I can help you... I can help you, I can, I can..."

Cia sighed.

"Please, don't just stand there... Talk to me... Let me hear your voice..."

"You said you could help me."

"Help you? With what? I'm sure I can. Of course I can."

"You said you could help us find land."

"Yes. Yes, I can. Oh, I can."

"Are you lying?"

"No! I can, I promise."

She sighed. Looked at the items in her hand. Were they going to be enough? Would a man like this not be able to outmuscle her?

But, again, did she have any choice?

"I'm not just going to let you out, Paul," she said. "There's going to be conditions."

"Okay. That's fine. I can do that. Conditions are okay, are good, so good..."

"You're not going to be walking around the ship freely. You're going to be bound, and I'm going to be in control of you."

He fell silent. Stared at her, his eyes bloodshot, wide like he was staring at a bright, shining light.

"I understand," he finally said.

His words sounded so sincere; they really did – it was his face that gave him away, that made him seem dishonest. He was desperate, and would say or do anything to get out – and that meant he couldn't be trusted.

Once again, her instincts screamed at her to find another way, but her mind said *no, I have no choice.*

"Stand against the far wall," Cia said.

Paul backed up to the other side of the cell.

Cia stepped forward, slid the metal stick with the noose between the bars of the window, and stepped back again.

"Tie the rope around your neck," she said.

"What? I'm not going to–"

"Do it, or I won't let you out."

With a huff, Paul picked up the metal rod, inspected it, then put the rope around his neck.

"Happy now?"

"Tighten it."

He bit his bottom lip, contained his fury, and did as he was told.

She threw the handcuffs through the gap, and took a step back again.

"Put these on."

He picked them up, and went to put them on in front of his waist.

"No, behind you," she said.

He tried to hide his glare. He failed. He evidently wasn't used to people telling him what to do, but he placed his hands behind his back then handcuffed them nonetheless.

"You happy now?"

"Let me see."

He turned around so she could see that the handcuffs were securely around his wrists.

"Now can you let me out?" he said, approaching the door.

"Stay against the wall," she instructed.

Reluctantly, he did as he was told.

"With your back to me."

He sneered, then he obeyed.

"And don't move until I tell you."

She approached the cell door, put the key in, and paused. This was it. No turning back. Whatever was in this cell, she was letting it out, and would have to deal with the consequences.

She turned the key. Opened the door. Stepped in, flinching as her foot accidentally nudged the remains of Paul's cell mate. The stink that had been apparent moments ago grew stronger; a hideous odour of faeces and rotting meat.

She took hold of the end of the metal stick, and, keeping her distance, she backed up.

"Okay, come on," she said, and guided him out of the cell.

CHAPTER FOURTEEN

Cia and Boy perched on the edge of their seats in (what used to be) the cafeteria, opposite Paul, not noticing how hard and uncomfortable the plastic seats were; they'd had far, far worse. Boy retrieved a tin of food from the store cupboard – steak pieces or something – and Paul scoffed it down like a dog.

"You seriously not going to let me use my hands?" he asked, lifting his head momentarily, gravy smeared across his cheeks.

Cia didn't flinch. She wasn't even considering the possibility of releasing him. She kept hold of the metal stick attached to the noose around his neck so she could keep him out of arm's reach, and refused to release his bound hands.

"I'm not letting you go," she said bluntly.

He grimaced and buried his head once more in the plate of uncooked juices. However foul his meal might be, Cia was sure it was better than what he'd previously been eating.

He finished the last few mouthfuls, twisted his head to the side, and wiped his mouth on his shoulder. The prison uniform was crusted with blood, and the lettering long since faded. She wondered if she should find him some new clothes, but realised that she'd have to take the handcuffs off for him to undress.

Cia looked at Boy. He was staring nervously at the prisoner, like he was on the verge of crying. She placed a hand on his knee and said, "It's okay."

"So what's your deal then?" Paul asked. "You two fucking or what?"

Cia frowned. "Excuse me?"

"Just wanted to know what was up with you two. You look a little out of place is all."

"What does that mean?"

"Well he's a lanky fucker, and looks like he's got some serious issues, and you're a little half-cast girl."

"I am not half-cast, I am mixed race."

Paul shrugged his shoulders and rolled his eyes. "Do I give a shit?"

She watched him eat the last few mouthfuls – watched the way his tattoos wrinkled as he moved; how every time his cracked lips parted she could see black between his yellow teeth; how his every mannerism reeked of arrogance.

He'd been locked in that cell for a long time, next to the corpse of a man he was forced to eat, unable to move more than a few steps. Doesn't that kind of thing drive a man mad?

And what if such a man is mad already – then what becomes of him?

"What were you in here for?" Cia said.

He snorted. "Lots of shit."

"Such as?"

"Why'd you want to know?"

"Because I looked through the prison files," she lied. "I didn't see anyone called Paul."

"Really? Out of every single man who was on here, there was no Paul?"

"No."

"Shit. I thought if I gave you a name like that, there was bound to be someone."

"So Paul's not even your real name?"

"At least I gave you a name. I don't even know yours."

She paused, then reluctantly said, "I'm Cia."

"Cia. And what's his name?"

She looked at Boy. She didn't know what his name was. She probably knew once, but had no reason to retain the information, so she ignored the question.

"So are you going to help us?" she asked.

Paul frowned. "With what?"

"You said you could help us find a way to shore."

He chuckled. "I did, didn't I?"

"Well? Can you help us?"

"Why should I?"

"Because I let you out – that was the deal!"

He grinned, and a dollop of saliva trickled down his chin.

"Or do you want to still be here when that creature comes back?" she asks.

"Honey, I've been here every single time it came back and rocked the boat. You think I'm bothered by it?"

Cia couldn't believe this. "Well maybe I should put you back in your cell."

"Maybe you should try."

"You think because I'm smaller than you, I can't?"

"What, you put these cuffs and noose on me because you think you can take me?"

She was already regretting what she knew she'd regret. This was a disgusting, vile, piggish man that she should have left to rot.

But she had to find a way to shore. They couldn't stay here.

She took out her knife and placed it on the table, letting its weight force an echo. She turned the knife so its handle was pointed at Paul.

"You may be bigger than me, but with your hands cuffed and that noose around your neck, who do you think would win?" She leant closer. "I've fought creatures far bigger and scarier than you, and I've won every time. So try it. Go on, I dare you."

His grin widened. "I'm starting to like you."

She raised her eyebrows expectantly.

"Fine," he said. "Take me to the helm. Let's see about getting us off this shitty little ship."

CHAPTER FIFTEEN

Cia kept Paul in front of her, guiding him with the stick along the deck of the boat, toward the cabin.

She could see how much he hated it, how humiliating it was, what damage a small girl being in control of him was doing to his ego. Every few steps he'd stumble a little, drag his feet, and would claim he wasn't used to being able to walk so much, but she knew he was trying to make things as difficult as he could for her.

When they finally reached the cabin, they had to shuffle in, stepping over a body as they did. She tried to keep her focus on Paul.

"Shit," Paul said, staring at the body. "Some fucker really did a number on this ship, didn't it?"

She shoved him further in with the stick, and he turned to glare at her.

"We're here," she said. "Go on then."

He looked from the metal stick to Cia's belt, where her knife resided. His eyes hovered, staring numbly, before looking back at Cia.

"Whatever you say," he said with a fake, demented smile, and turned to the helm, looking at all the screens and buttons around it.

"Can you see a map?" he said.

Cia looked around, and found one behind her. She placed it beneath the window for him. He glanced at the radar, then back to the map.

He kept doing this, before turning to the buttons, and looking at what was written beneath each of them. Studying everything like he was seeing it for the first time, trying to learn what everything did. He kept referring to the map, like it was for show, then turning back to the radar, then the buttons.

Did he actually know what he was doing?

She watched him do this for a while, eventually turning to Boy beside her, whose look of terror seemed to be permanently etched to his face.

"You don't know how to get us to shore, do you?" Cia said.

"Fuck off and let me concentrate."

"You're blagging it."

"Fuck off, I said!"

He continued to look over all the knobs and dials and buttons, clueless as to what he was looking for.

This was when Cia realised just how much of a mistake they'd made. He'd just said whatever he had to, to get out of that cell. He'd have told them he could walk on water if they'd believed him. He'd predicted what they needed, anticipated what they feared, and played on it.

And, as Cia watched him, she also realised that she was going to have to undo her mistake.

She was going to have to put him back in his cell.

She dismissed the thought. He'd die before he let her drag him back to those claustrophobic walls.

Which left one more solution.

She was going to have to kill him.

She had killed before. Against her better wishes, and only when she needed to, yet the trauma still trickled through her thoughts, still conjured images every time she closed her eyes – but she had to protect Boy and herself. This man was too dangerous; she could see that now.

She could never let him go free of those handcuffs. She was never going to be able to let him roam around the ship. And he was never going to be able to get them back to shore.

She had to do it. Now, while he was busy pretending to know what he was doing. While he was figuring out the ship's controls so he could continue this charade.

She gripped the handle of her knife. Took it out.

Not another one.

She really didn't want to.

How many is that now?

It was a different world.

We do what we need to survive.

But why did it always come down to this? Her survival pitted against the sadistic nature of the human race?

Stop thinking. Just do it.

She met Boy's eyes, and put a finger over her lips, shushing him.

She flexed her fingers around the knife. Tightened her grip. Edged toward Paul's back.

If he notices he'll try to kill me...

She willed the thoughts away. Stop thinking them. Can think later. For now, just focus. Don't ruminate, don't analyse, don't consider what that knife is going to do – *just do it, and think later.*

She raised the knife, getting ready to pounce.

Then something beeped.

She instinctively put the knife back to her belt as Paul turned, drawn toward the beeping.

The beeping grew more frequent.

It was coming from the radar.

"What is it?"

Paul looked at the radar. "Something's coming."

Cia stepped forward, looking at it too. A line spun around its green screen, revealing an object, something large, and each time the line passed it, it was closer.

"What is that?" she asked again.

He didn't answer.

"Is that a ship?"

"That's no ship."

"How do you know?"

"It's too big to be a ship. It's not moving like a ship. That's something else."

His gaze lifted to the window, and he stared into the distance.

The ship rumbled, the floor vibrated beneath her, the light above them swayed.

She gripped Boy, digging her fingers into his arm, unknowingly dropping the metal stick.

A large wave pushed the frontend of the ship into the air, where it paused before smacking back to the ocean's surface, water crashing over the side.

She steadied herself with the doorway and peered into the distance. The water rose as a lump beneath the sea grew closer.

The creature reached the ship. Knocked into its side. It almost capsized, then rebalanced, sending them against one wall, then to another, then back to the ground.

Then Cia saw it. Coming out of the sea. Its deep murmur booming through her body, rattling her bones, its large head rising over the ship, its various tentacles waving, huge, so huge.

It had come back last night, searching for survivors.

And now it had found them.

CHAPTER SIXTEEN

Boy.

That was Cia's first thought: Boy.

He couldn't swim. Him falling overboard would be as much of a death sentence as contending with this creature. Whatever happened, she had to keep him on the ship.

But would there even be a ship left if this monster kept hitting it?

It was hideous. Horrifying. Bigger than a Thoral and more terrifying than a Lisker. It was a sea creature, with a squid-like appearance, and she was sure she'd be safe from it should they be on land – but they were not. They were in the middle of the ocean, and its tentacles were flailing about, searching for its prey, and its eyes – wide slits in its huge, domed head – were looking right at her.

"Come on," Cia urged Boy, grabbing his hand and pulling him out of the cabin.

It hadn't destroyed the ship yet, and Cia held onto that thought. She didn't understand why, considering it could do so with ease, but it hadn't. So if they buried themselves inside the ship, deep inside, and waited, and endured the constant panic inspired by its knocking the boat from side to side, could they survive?

Of course, if it knocked the ship over, they'd be submerged in water, stranded in the ocean and left to drown – but what else could they do?

They ran across the ship's deck, past the windows, a door in sight but out of reach. Paul's chaotic footsteps followed them, but she didn't care – he could do what he wanted. His safety wasn't her concern.

Then came another swipe of a tentacle, pounding into the side of the ship, sending all of them falling onto their backs. Another tentacle slammed onto the deck, smashing part of an adjacent room.

Cia pushed herself to her feet and dragged Boy back the way they came. She glanced over her shoulder, catching sight of its large head blocking out the sun, its eyes red and narrowed, and she pulled Boy to the side, hopefully toward a room, but definitely out of its sight.

It knocked into the ship again, rocking it. Waves crashed over the side and sent them stumbling back to the floor. Cia tried to keep hold of Boy's hand, but it was too wet. He slid out of her grasp and she could do nothing but watch as he slid further down the sloping deck.

"No!"

She slid after him, until another smash tilted the ship the other way, and Boy was sent hurtling back toward her.

He knocked into her, escaped her grasp again, and slid toward the railings. He managed to grab hold of one, and clutched with all he had as his feet dangled over the edge.

The ocean reached for him. One moment she could see him screaming for her, the next he was submerged in a ferocious wave – which fell away quickly, leaving Boy soaked and bedraggled, crying as he hung on.

Cia leapt across the deck, clinging onto the railings, reaching for Boy. His outstretched fingers were inches away. As she stretched, almost touching his hand, another tentacle pounded into the ship and rocked it once again. Her ankle became entangled in a loose bit of rope, and she was pulled away from the person she loved most in all the world.

"Hold on!" she shouted, desperate to be heard above the chaos of the sea.

Boy's lip was quivering. His arms shook. The moisture on the rails was forcing his hand to slip, and he was struggling to hold on, and it was the only thing Cia could focus on.

Behind her, Paul was shouting, his legs wrapped around the railings, but she didn't look over her shoulder – she remained focussed on Boy.

Again, she tried to reach Boy.

Again, the rope held her back.

Should she disentangle herself then try? Or would it be too late?

Could she still reach him? She was determined. Stretching her arm. His fingers reaching out.

But it was too late. Boy was out of reach, his hand was slipping, and as another smash of the tentacle rocked the precariously balanced ship, Boy's grip loosened and he was forced to let go.

Cia could do nothing but scream as she watched him plummet into the depths of the ocean.

CHAPTER SEVENTEEN

Cia dangled upside down from the railings, the blood rushing to her head, struggling to think, adrenaline taking over. She stretched her arm up, reaching for the rope wrapped around her ankle.

But her fingers could only just brush it, and she hadn't the strength to hold her body up.

She looked over her shoulder. She could see Boy, small and screaming, thrashing his arms as the ocean carried him away. The creature's multiple thrashing limbs were sending waves toward him, and he was scrambling to stay above the water's surface.

She refused to cry. And she refused to give in.

After all that she had protected him from, she wasn't going to watch him drown. That was *not* how she was going to see him go.

She reached for the cables, and failed again.

"Give me the keys!" came a voice from beside her.

Paul hung upside down, his legs wrapped around the railings, the noose still around his neck and his hands still bound behind his back.

"Give me the keys and I'll get him!"

She looked into Paul's eyes, the eyes of a mad man, knowing they could not be trusted, but not caring one bit if it meant he would help Boy.

She reached into her pocket, grateful to find the keys still there, and stretched her arm out to Paul, who took them in his mouth. He threw himself back onto the deck, twisted his body, spat them into his hands, and released himself from the cuffs, throwing the noose into the water as soon as he freed his hands.

Then he looked at her.

And he grinned.

"Save him!" she said. "Please."

He laughed at her, gave her a wave, then ran across the ship.

"You bastard!" she screamed.

Another smash of a tentacle knocked the ship, forcing Paul onto his back, and he slid across the deck. His smile quickly faded as he fell through the crack of the railings and into the ocean.

It was as much as he deserved.

The latest knock into the side of the ship had made it level once again, and Cia could now twist to the side, closer to the rope. Fleetingly appreciating her small bit of luck, she stretched her arm out, further and further, reaching with all she had until she could dig her fingers between the pieces of rope, releasing herself and diving into the water without a second thought.

She swam hard, her muscles struggling against the aching; her body resisted but she ignored it, and surged toward Boy.

"Boy!" she shouted, knowing he would not hear her above the pounding of the waves, but trying anyway.

The waves shoved her back and forth, to and fro, under the water and back up again. She plunged herself under the surface to get a better look and, as the water pushed her back, she saw him. He wasn't too far. He was underwater, but then he returned to the surface, so she knew he could breathe. She just had to get to him.

She pushed her arms, kicked her feet, willed herself forward, using every bit of resolve and determination she had – her body was struggling against fatigue, but it was her mind she had to convince.

She promised herself she had the energy. The vigour. That she could do this. She rose above the surface, and she could see him, nearby.

"Boy!"

A large wave dragged her further away and she could do nothing to resist its force.

He turned to look at her.

"Rosy!" he shouted back, his face a contortion of terror.

She looked over her shoulder. The creature appeared confused, as if it had lost them, like it was trying to figure out where they were. It thrashed its tentacle into the water, an undeniable act of frustration that sent waves toward Cia and carried her toward Boy.

She used its momentum to push forward, swimming furiously.

Another frustrated smash of a tentacle created more waves and sent her closer still.

She kicked her legs. Moved her arms. Swam with everything she had. Pushed herself with the little energy she could find until, eventually, *finally*, she reached his side, and was able to wrap her arms around him as they bobbed on the seabed.

She was grateful, but it still wasn't over. They needed to get back on the ship, and they needed to do so before the creature realised its meal was waiting for it, helpless and terrified.

CHAPTER EIGHTEEN

Where was Paul?

She didn't care for his wellbeing, not after what he'd done – but she cared whether the creature was distracted. If it was chasing Paul, that would give her the opportunity she needed to climb back aboard without it noticing.

The ship wasn't far, but it was taking too long – she wasn't running toward it, she was swimming, fighting fatigue, and dragging Boy as she did.

The waves calmed. She wondered if the creature had found Paul, and if it was feeding, and was therefore distracted.

She didn't care. She didn't need an explanation as to why she was able to escape, she just needed to do so.

Even so, it was odd how the sea could go from furious to lethargic so quickly; how the creature was there one moment, and now had seemingly disappeared.

In Cia's experience, if something was too good to be true, it often was.

They reached frontend of the ship, climbed the ladder,

and collapsed on the deck. Cia looked over her shoulder, searching for their attacker.

Where was it? Where had it gone?

She wanted to know – she needed to know. But she couldn't allow herself the luxury of waiting to find out.

"This way," she told Boy, and she led him into the first door she found.

It was a kitchen. All the plates were on the floor, smashed, along with the glasses and cutlery. Bodies still remained, but the tilting of the ship had changed their position. Some were against the wall. Some were caught on a table. Some piled on top of each other. She didn't stay to look – she stepped over them, quickly, forcing Boy to have to run to keep up.

They reached stairs. Cia guided them down a few floors, deeper into the ship. Through a corridor, then another.

She passed a cupboard with a few vents in the door, and paused. Opened it. Grabbed a few towels. Carried on running, glancing in the windows of each room.

Most contained corpses. As soon as she found one that didn't, she led Boy inside and closed the door behind them, as if shutting it would do anything to protect them.

Then she stopped.

Stayed still for a moment, and felt it all catch up with her.

She leant against the wall, panting so hard she was beginning to wheeze. Her heart thud against her chest. Muscles throbbed. Her clothes were wet and heavy. She just wanted to lay down and fall unconscious.

She looked at Boy. He was much the same. Drenched, panting, shivering. Like her, the water had left him frozen cold. He was sobbing. She didn't blame him. She wished that she could indulge herself in a few tears, could allow herself to show just how scared she was, and how horrifying the ordeal had been.

But she couldn't. She was the protector. Not only did she need to keep Boy safe, she also needed to keep him calm; if he shut down, refused to move or listen, then there was little she could do to bring him back. He was in desperate need of her strength, so she wrapped her arms around him.

She had to be impervious to the pain, at least in front of Boy, and save the sobs for when he was asleep.

"It's okay," she whispered into his ear.

She was waiting for another shake of the ship, for another attack, or something – but it didn't come. The creature had stopped. She didn't understand why, but it didn't matter. Normally these creatures were persistent, and she was surprised it wasn't tearing this ship apart; though she was also grateful.

If it wanted, it could tear the ship to pieces and sink it to the seabed.

So why didn't it?

Shut up, Cia. Enough already.

These questions could come later.

If at all.

It wasn't like she'd understand the actions of a monster.

Her gaze abruptly shifted to the window. She thought she'd seen something.

She hadn't. There was nothing there.

She was seeing things.

All this trauma, it was bound to screw her up. She was surprised her imagination didn't trick her more.

"Come on," she said. "Let's get you dry."

She took the towels she'd collected and began drying Boy as he undressed. He took over, and she took a towel, undressed, and dried herself.

As she did, she noticed something on Boy's bicep. A large wound.

"What's that?" she said, and Boy looked at it then panicked.

She grabbed his hand and said, "It's okay, it's okay. We'll find a medical room or something and clean it up."

She was surprised he hadn't noticed the pain. Then again, the adrenaline and the terror were probably numbing it. Once he calmed down, that's when it would really sting.

Noticing he was sobbing again, she pulled him close and hugged him.

At least she could be comforted in the knowledge that it was very, very unlikely that Paul would have survived.

CHAPTER NINETEEN

A HAND REACHED high above its body and gripped the damp railings, flexing its fingers around it. The second hand joined it; the words FUCK and LOVE clearly visible tattooed along the knuckles.

With strength only gained through boredom, from hours spent in the gym because prisons leave you with sod-all else to do, he pulled himself up. Once his feet were high enough, he placed them against railings and climbed over, falling onto ship's deck with an *oomph*.

From afar, they were visible. The girl and the boy, climbing aboard and rushing into the corridor. With a growl, he spoke the words aloud as if it was the only way to make him believe it: "They survived."

How the hell had they survived?

He looked over his shoulder. He did not meet the creature's eyes, though he knew they were there. Watching him. It had stopped attacking when it lost sight of the boy and girl, and its anger was evident.

Paul had failed, and he knew the penalty should he do so.

But he never failed.

Never.

He'd let them put handcuffs on him. He'd let her drag him around on a stick. He'd suffered humiliation to play along with his façade, to make them believe he was subservient; as if he would ever actually let anyone treat him like that.

A prison officer tried humiliating him once. Tried to show that Paul couldn't be rude to him by knocking his lunch over. Paul had knocked three of his teeth out and broken his jaw in a single punch.

He turned, not daring to make eye contact with the creature, and said over his shoulder, "Give me a day."

He paused. Waited. Closed his eyes, flinched, and prepared himself.

When he opened his eyes, it was gone, and he was still alive. That was as much confirmation as he needed.

He charged ahead, marching forward until he reached the door the boy and girl had gone through, and entered the kitchen, then the stairs. It didn't take much effort to follow their soggy footprints. They took him down a few floors, along another corridor, and to a door he could easily open.

He stared at the door, trying to be calm, considering his next move. He couldn't fuck this up again. He had to be methodical, had to think carefully about this.

Yes, he could just barge the door open and grab them. He could drag them out, both of them, and throw them on the deck. But he had a feeling this girl wasn't as weak as she looked; at least not when it came to brains. If they had survived for this long, then she would have fought through some horrendous situations to get here. If she managed to break free of his grasp, or managed to flee without him grabbing her, they would probably hide and he'd never find them within the day he'd promised to the creature.

He had to be cunning.

He had to be smart.

He needed to lure them out, and it was pretty clear how.

He approached the room. Paused at its edge. Watched them through the window. She was hugging the boy.

No, not just hugging...

She was *clinging* onto him. She was squeezing him, pressing their bodies together, holding onto him for dear life.

He was her weakness.

On his own, he was easy to manipulate. She was the one who protected him, and she would do so with everything she had.

The boy was how he'd get them.

She turned her head toward him, evidently sensing he was there, and he quickly backed up.

He waited to see if she came out, but she didn't. He found another room, somewhere he could think; somewhere he could formulate his plan.

The ship was steady again and the ocean was calm. The fight was over.

At least, as far as the boy and girl were concerned.

If anything, the real fight had only just begun.

BEFORE

CHAPTER TWENTY

THE WALLS of Paul's cell were closing in on him. It was impossible to think. He'd been trapped in this six-by-eight-foot room with nothing but a bunk bed, a toilet that no longer flushed, and the half-eaten corpse of his cell mate.

He'd been diagnosed many times with many mental illnesses. He'd never paid any attention to it; they could say he was mad over and over, but he knew the truth. He was just smart enough not to betray his true nature.

Yet, as he sat in the middle of the floor, rocking back and forth, accustomed to the smell of shit from the bog and the rotting of the dead man, he could feel himself slipping further into insanity, his thoughts becoming more erratic, no longer in control, sliding away from him like two bloody hands clinging onto a fading memory.

"You..." he grumbled, turning to his cellmate. He'd long since forgotten his name. "You... did this... to me..."

He glared at the man whose head remained, despite most of his body having been eaten. His eyes were wide open, stiff and staring.

"You... fucker..."

With a roar he leapt across the cell and pounded his fist into his dinner's face, over and over, until he remembered the man was dead and could feel no pain and he toppled over with laughter, chuckling and giggling and snorting and hooting.

The guy was dead!

Hah!

He wouldn't be able to know he was a prick however much Paul told him.

"You… idiot…"

Who was he talking to?

Himself?

His cellmate?

The voices that told him all those nasty things?

Footsteps.

He sat up quickly, his breath catching, suddenly serious.

Footsteps came down the corridor. Someone was there.

"Hello!" Paul shouted.

The footsteps sped up.

"Hello, I'm here!"

The footsteps quickened. They were running toward him.

By the time he'd pushed the remains of his cellmate under the bed, the footsteps had stopped and a face was at the door. It was a man with a large beard and friendly eyes.

"Oh my God, are you alive?" the man said.

"Yes," Paul said, finding this a stupid question.

"Jesus, we thought we were the only ones."

"We?"

"Yes, there's only a few of us that survived."

Paul wanted to ask what it was they'd survived, but for some reason, he didn't.

"Let me find the key," the man said.

Metal jangled as he searched through a large set of keys. He'd try one, find it didn't fit or didn't turn, then try another.

Paul watched him, wondering what his insides looked like.

"I'm Graham, by the way," he said.

Paul decided to go by his middle name. He didn't want Graham to hear his first name and know who he was.

Graham probably wouldn't let him out if he knew who he was.

"My name is Paul," he said.

"Hi Paul," Graham said, sounding flustered as he tried yet another key.

Finally, this one worked, and the door opened.

Paul stepped out.

Wow. This was what freedom was like.

He extended his arms, feeling the open space. He kicked his legs and stretched them, allowing his body to walk.

"Come on," Graham said. "Let's get you some food."

Graham walked on and Paul followed, unable to help himself from grinning.

Funnily enough, Paul wasn't hungry.

As Graham led him out of the corridor and to the stairs, Paul noticed a knife attached to Graham's belt.

They reached the deck, and Paul took a large intake of clean, ocean air. He hadn't breathed in air this good in so long.

All around them was ocean, but the ship was big, and he no longer felt trapped. Moisture clung to the air, and the sound of gentle waves pushing against the ship provided a pleasant soundtrack to his liberation.

"The rest of us are in the cafeteria, it's this way," Graham said.

Almost as soon as he said it, something changed. The

calm ocean became a little less calm. In the distance, a large wave came closer.

"Oh, damn," Graham said. "It's coming."

It was coming?

What was coming?

"Quickly, this way," Graham said, rushing along the deck.

Paul did not rush. He let Graham run, but just paced behind, too curious as to what was approaching.

It took seconds to arrive and burst above the surface, its large head surrounded by thrashing tentacles.

"Shit," Paul muttered, a little out of fear, but mostly out of amazement. It was a thing of beautiful destruction.

"Come on!" Graham said. "We need to get inside!"

But it was too late for them to get inside, and they both knew it.

The monster swung its tentacle against the side of the ship, sending Graham onto his back. Paul, however, had enough sense to cling onto some railing before the tentacle hit them.

Then it looked at them. Both of them. Graham and Paul.

It looked at them and opened its mouth, ready to devour whichever it collected first.

Paul wasn't prepared for it to be him.

He ran to Graham's side and took the knife from his belt.

"What are you doing?" Graham asked, and Paul saw it in his eyes – that flicker of fear, the one that each of his victims had just before he snatched their life away.

It was a moment in which they realised they were going to die.

Paul slit the knife across the back of Graham's thighs, forcing him to wail as he collapsed.

Graham tried to stand, but the pain was too much.

Paul turned and ran back the way he came.

By the time he reached the door, Graham was dead, and he'd escaped with the knowledge that there were more survivors.

More sacrifices.

More people to give the creature to avoid his life being taken.

It couldn't have been more perfect.

AFTER

CHAPTER TWENTY-ONE

Cia wrapped what little bandages she could find in the first aid kit around Boy's arm, covering the large cut on his bicep. She'd hoped she could spare some for the wound on her navel, but could only spare enough for Boy.

It still wasn't enough. They were going to need painkillers, maybe even antibiotics.

"Come on," she said. "Let's go see what we can find."

She edged to the doorway, looking back and forth down the corridor, and stepped out. She kept her hand in Boy's as they walked, providing him with the reassurance she knew he so desperately needed. He was doing well, but she knew it wouldn't take much to change that.

They left the floor and descended the stairs, passing a few more corridors of cells, until they came to a corridor that wasn't just lined with heavy cell doors, but also with offices and storage rooms – perhaps this floor was more likely to have what they needed.

She entered, peering inside the first few rooms. Just offices. Decaying bodies, ripped paper and upturned furniture wasn't much use to her.

She continued through the corridor until she reached a small room with the word *Armoury* on the door. Not quite able to believe she'd stumbled on a cupboard full of weapons, she pushed the door and looked inside. Sure enough, there they were. Shelves of guns and ammunition.

Cia hadn't used guns before. She'd used knives, and she'd used her wits – but she'd never seen the advantage in firearms. A gun was difficult to acquire in the UK without knowing where to look – and even if she did know where to look, they were noisy, and would probably do little damage to the creatures whilst attracting more of them.

Even so, it wouldn't hurt to take one. It would be better to have a gun and not need it, than need a gun and not have it.

On the far wall were shotguns and rifles. To her right were boxes of ammunition. To her left were smaller handguns, which is what she opted for. She was better off using one of these, considering her lack of experience.

She took one. It said *Glock* .57 on the side, so she searched the boxes of ammunition for the appropriate bullets, and found one, along with another magazine full of bullets.

She pressed a few buttons on the side, and it released the barrel. As far as she could see, it had bullets in, so she placed it back in.

She marvelled once again at the cupboard before she left it. Every shelf was full, except for the space she'd left by taking the Glock. It would definitely be worth revisiting this.

"Rosy..." Boy moaned, grabbing his arm. "It's hurting..."

"Oh, I'm sorry," she said, feeling bad for being distracted. With the adrenaline fading and the bandage pressing against his skin, she had no doubt Boy's wound was stinging, so she closed the door to the armoury and continued through the corridor until they reached a medical room. From it, she took a few boxes of painkillers, antibiotics, and a bottle of water, then carried on through the corridor, looking for a room vacant of corpses.

She came across a room labelled *Records* and led Boy inside. There were a few cabinets against a wall, a few chairs, and a desk. She sat him on one of the chairs, then sat opposite him and sifted through the various antibiotics.

She found one called Amoxil, and checked the label. It seemed okay to take, so she gave them to Boy, along with the water and two paracetamols, and watched him swallow them.

"Okay?" she said.

"It still hurts."

"It's going to take a while," she said. "Just wait for them to kick in."

He moved from his seat to the one beside Cia and rested his head on her shoulder.

"Try to get some sleep," Cia said, dreading the thought of spending another night on the boat. When she had woken up that morning, she had been determined that they would find land. Instead, she had released a criminal who posed as much threat to them as the monsters, and were no closer to leaving this ship.

Cia leant her head back and closed her eyes.

After a few minutes, Boy's breathing became deeper, which led into a gentle snore.

But Cia couldn't sleep.

She couldn't stop thinking about Paul.

There was no way he could have survived the creature's attack.

Was there?

She hadn't seen his body – but if the creature had eaten him, then she wouldn't, would she?

She lifted her head. Opened her eyes. Sighed. This was no good. Things were playing on her mind too much.

Her eyes wandered around the room, past a few posters about fire safety and emergency procedure, past a few dents in the wall, and to a cabinet that was nailed to the ground.

Didn't the sign on the door indicate that this is a room of records?

Did that mean the cabinet held information about Paul? Could she find out who he really was, and what he was capable of?

She gently took Boy's head off her shoulder and lay him across the chairs. He groaned a little, but didn't stir, and she stroked his hair back and placed a kiss on his forehead.

She approached the cabinet and opened the first drawer. It contained many, many files, all organised alphabetically by surname.

She didn't know Paul's surname.

Hell, she didn't even know if Paul was his real name.

Screw it, it wasn't like she was running out of time. She took a load of files, placed them on the desk and looked through them. Each had the prisoner's name, picture, and list of convictions.

Mohammed Ahmed.

Simon Allen.

Kamal Quentin Atkins.

This was going to take a while...

She forced her aching arm to keep moving from one file to the next, and for her fatigued yet frantically anxious mind to focus. She did the As, the Bs, the Cs, and it wasn't until the Ks that she was really starting to lag.

She stood. Straightened her back. Wiped sweat from her forehead.

Looked at Boy, who remained soundly asleep.

Sighed.

Continued.

Ls. Ms. Ns. Os.

None of the pictures looked familiar.

There were plenty of people who'd done horrific things. Murderers, paedophiles, drug dealers, traffickers – all kinds of people aboard this ship.

Ps. Qs. Rs.

She opened the final drawer and took out the Ss.

And, just as she was considering giving up, she found a picture she recognised.

A picture of Trevor Paul Scott.

She opened the file, looked through his various misdemeanours, and instantly wished she hadn't.

He was in prison for multiple counts of murder. He'd killed his wife and children. He'd stalked women on a local college campus and butchered them in the back of his car. He'd lured young boys away from schools during lunch time, and the bodies were found days later in the woods – mutilated and sodomised.

She realised she'd heard the name. She'd been far too young to understand this kind of thing at the time, but she had a glimpse of a memory of her father watching the news and switching it off as she entered the room because the headlines were about a serial killer.

This was a nasty, horrific man. He'd killed and abused girls and boys their ages. Even slaughtered his own family.

Why the hell had she let him out?

She closed his file. She couldn't read anymore.

She lay behind Boy and placed an arm around his waist, feeling even more protective of him. Boy had endured some dreadful experiences whilst fleeing these monsters, but she had still managed to protect him from the horrors she'd experienced.

It was *her* who'd had to kill a man whilst he penetrated her.

It was *her* who'd had to kill Dalton whilst Boy hid.

It was *her* who'd had to fight the evil that was left in the form of human nature.

She could never allow Boy to endure the awful experience a man like Paul could put him through. She would die before that happened.

But Paul must be gone. The creature must have killed him.

Still, she wished she'd seen a body, or had seen him die; something that would reassure that nagging doubt in the back of her mind.

No. Stop it. There was no way he could have survived.

He must be dead.

He must be.

She willed the thoughts away as tiredness overcame her and she fell into a light sleep. She didn't stay asleep for long, however, and ended up lying awake most of the night, listening to the noises.

Every light crash of a wave.

Every pipe.

Every creak of a battered ship.

Just waiting for footsteps.

Waiting for anything that would tell her they weren't alone on this ship.

But when morning came, they were still intact and undisturbed.

She told herself to stop being silly, and to try and sleep for another hour or so. They were alive, and they were alone. There was no one else on this ship.

No one at all.

CHAPTER TWENTY-TWO

Cia was on the floor. Paul was on top. Reaching at her clothes, tearing at her flesh, his grin wide enough to release droplets of drool – droplets that crashed upon her chin and splashed upon her lips.

She writhed and wriggled and even spat, but his smirk only intensified and the lump against her waist only grew harder.

"Please, stop it!"

She had never felt more helpless.

How did she let him do this to her?

She'd let him out. It was her fault.

But he was dead now.

She was sure of it.

How was he not dead?

His slimy tongue met her neck, flicking up and down. She flinched away, scrunched up her face, refused to cry or yell or shout or ever stop fighting.

She would resist until he killed her.

Then she remembered... Boy.

If he killed her, would he be next?

He liked boys too. The report said as such. Young women and boys, that was his demographic.

So she stopped resisting.

She lay flat, stiff, unyielding; not fighting but not obeying either.

But she didn't need to be subservient. It only took a few of his odious breaths for him to remove her clothes place himself inside of her.

He thrust and it hurt and she screamed and, finally, she woke up.

Sweating.

She leapt to her feet and looked around the room.

Boy began to stir, having been laid on the chairs next to her.

"Rosy?" he said, rubbing his eyes.

"Shush," she said. "Go back to sleep. I'm fine."

With a sleepy groan, he lay back down and closed his eyes. Cia did not stop panting. She placed her hands on the desk, spreading her fingers out. She could still feel him inside of her. Even though it wasn't real, she could still feel it, like a rod, plunging deeper and deeper.

She needed the toilet.

And, unlike most of the time she'd spent in this world, there was actually a real toilet down the corridor for her to use. One forgets how much they never used to appreciate such a luxury.

She took the gun and the keys and crept out of the room, locking the door behind her, and made her way down the hall and into the bathroom.

The light of dawn came through the window, illuminating the room in gentle shadows. The mirrors were cracked, but she was still able to see her reflection – something she hadn't seen in a long time. Her hair was bushy and unkempt. Her teeth were dirty. Her skin had splotches of stains she couldn't identify – but she was alive, and the reflection proved that at least.

She went to open the first cubicle, then stopped as she saw a few maggots crawling over the feet of a dead woman sat on the toilet, and went to the next cubicle.

She sat down, the porcelain a little chilly on her thighs, and considered, as she used the toilet, about mental health. It was a concept they used to talk about before all this happened. It was even something her father had discussed with her.

Now it was something she couldn't afford to contemplate.

And that's when she realised – she was pissing in a cubicle next to a dead woman, and she hadn't even given it another thought. Was that what her life had become? She was so used to death that she could be in the same room as a corpse and barely even notice?

She finished. Buttoned up her trousers. Stepped out. Leant over the sink.

And she cried.

She'd had enough of fighting. She'd had enough of killing; both monsters and humans. She'd had enough of worrying whether Boy was going to have a breakdown whilst she tried to save his life. She couldn't take another minute of having no other purpose than to live another day.

Was that really a life she wanted to live?

And, for a moment – just a matter of seconds, in fact – she gave up. All the trauma became too heavy, and she sobbed and sobbed, and decided she may as well die because this life was not worth it.

Then she thought of Boy.

Of how she'd feel if she lost him.

And she no longer felt like giving up. She no longer felt like she had no purpose. She no longer felt like all of this was pointless.

She stopped crying. Took a deep breath. Looked herself menacingly in the eyes, and strode out of the room, eager to find a way off this boat.

She'd only taken a few steps when something caught her attention.

The armoury.

The door was open.

She was sure she'd closed it...

She scanned all the shelves. There was the vacant space where she'd taken a gun, and... there was another vacant space further along from it.

"Oh god..."

Someone had been here.

Someone had taken a gun.

"Boy..."

She ran, skidding to a halt as she saw the open door to the room they'd been sleeping in.

"Boy!"

She pointed her gun at the doorway as she approached. Waiting for something. Anything.

Boy was no longer lying on the chairs.

Boy was not at the desk. Or at the other side of the room. Or by the filing cabinet.

Boy was not in here.

"No..."

She turned and shouted, so hard it broke her voice and reverberated against the corridor walls over and over.

"*Boy! Where are you!*"

She listened.

Nothing.

She tried not to be sick. Tried to ignore the churning of her belly, the shaking of her arms.

Please, say he just wandered off... Say Paul didn't come back and take him... Please...

"*Boy!*" she tried again.

And, after a few unsettling seconds, she heard him, faint yet definite, shouting back in terror.

She sprinted in the direction of the voice.

CHAPTER TWENTY-THREE

Cia hunted.

Like a Thoral would hunt her, or a Maskete waiting to swoop down and collect its winnings – she listened, and moved toward the sound.

"Boy! Keep talking to me!" she shouted.

His screams responded, faint, in the distance.

She no longer had any doubt that Paul had survived. That feeling in her gut that had been pestering her, that constant dread – it was telling her all she needed to know.

Boy kept screaming. Paul could easily cover Boy's mouth, stifle his words, or, hell, even kill him – but he hadn't. He was allowing Boy to respond. It could be out of some sick pleasure he gained from it, but, most likely, he was luring her in.

Paul wanted Cia to follow.

Which begged the question – where was Paul leading her?

And her answer: *I do not care.*

It could be to the very depths of Hell and she'd go there.

Hell would be nothing compared to what she'd already endured for him.

"Boy, I need to hear you!"

More screams. Still faint, but getting louder.

She reached the stairs. Opened the door. Listened – were they going up or down?

Footsteps beat against the metallic surface. She looked up and saw the shadows leave through the door to the deck.

She leapt up the steps, two or three at a time, clutching the gun she was dreading having to use; not because she was afraid to kill Paul, as she certainly was not – but because she would probably only get one chance to shoot Paul, and she could not screw it up. If he saw how little she knew about her weapon, he would gain a huge advantage.

She barged open the door to the deck and held her gun out, like they did in the movies she used to watch with her dad. Such memories were barely a glimpse now, but she recalled a flash from a movie, possibly James Bond, where he was walking through corridors with a gun, and he held one hand under the gun and the other hand on the grip, staring down the barrel at whatever was before him – so she did the same.

No one was there when she emerged onto the deck. She looked one way, then the other.

"Boy!"

"Rosy!"

His shouts were clearer now. She turned to her left, and headed toward the frontend of the ship.

That was where she found them. Boy on his knees, crying furiously, his head buried beneath his hands, moaning, rocking back and forth, repeating "Rosy, Rosy, Rosy," over and over.

He'd broken down. It would take a great effort from Cia to bring him back from this.

Behind Boy stood Paul. That same smug leer spread from cheek to cheek. A gun pointed at Boy's head.

"Drop the gun," he told her, his voice calm.

"No," Cia said.

Cia would have obeyed every word if she thought it was in Boy's best interest – but the same thought spun around her mind: *Why hasn't he killed him already?*

He wanted them alive. She couldn't understand why, but it was the only reason they weren't yet dead.

"Drop it or I'll shoot him."

Cia aimed at his head. She'd never aimed a gun before, but instinct told her to align the end of the barrel with his face and hope for the best.

Her finger stroked the trigger.

"I said drop it."

She pulled the trigger.

She waited for the kick back, expecting the gun to pack a punch as it fired.

It did not.

In fact, the trigger didn't even move, no matter how much pressure she put on it.

She ignored Paul's hysterics and examined the gun, searching for the fault. There was no safety button or anything on the side.

"It's on the trigger," Paul said.

Cia looked at the trigger. Sure enough, there was a latch on it. She slid her thumb over it, removed the safety, and felt stupid for not knowing something she could not have known.

"You lift that gun again and I'll shoot the kid before you manage."

She looked up at him, but left the gun at her side.

Boy continued to rock, moaning, covering his ears. To them, there was no sound but the waves – but to Boy, there were a thousand screams in his mind prompted by a frenzied confusion about what was going on.

"What do you want?" Cia asked.

Paul looked to the ocean, scanning the water, looking for something.

"Just wait," he said. "And you'll see."

Cia frowned. "See what?"

"I'm sure it will be here in a–"

Paul paused.

His smile quickly faded, replaced with a look of horror.

"No..." he mumbled. "No... You're meant to wait til after... It's not your turn..."

Cia wondered what he was on about – then she heard it too. A sinister crescendo of tapping, like a thousand claws against metal.

It grew louder.

Whatever it was, it was coming up the side of the ship.

"What is that?" she asked.

Paul didn't even hear her. He was too busy staring wide-eyed at the side of the boat, becoming increasingly alarmed with every second that passed.

"What is it?" Cia repeated, louder.

He looked at her, and she saw pure terror in his eyes.

He released Boy and ran.

Cia sprinted to Boy's side, took hold of his arms, and went to speak to him – but the sound was too loud.

She looked over the side of the ship and froze, unable to comprehend what it was she was seeing.

CHAPTER TWENTY-FOUR

They came from the depths of the sea.

Deep, deep below the surface.

At the bottom where the bodies lie.

They came and they rose up.

The larger monster always has the first chance to feed, see. That's how it was when the ship was previously massacred. The bigger monster took what it wanted then left the rest to them.

And, with the commotion of the previous day over, these creatures were sure there were remnants – further bodies to feed upon.

So they floated to the surface. All of them. A mass of creatures, no bigger than a metre, but as deadly as any Maskete or Thoral you might face. Six legs, three either side of its body. No eyes – it didn't need them; it could hear you. And where its belly would be, between its legs, was its mouth. A long slit with sharp fangs able to cut through bone like it was paper.

Hundreds of them.

Maybe even thousands.

And they didn't want to eat you; they wanted to *devour* you. They wanted to latch onto a limb, shred it apart, then slowly feast upon the leftovers.

They reached the ship, sticking their legs into its side and crawling up, up, onto the deck, scuttling in their masses, spreading out.

They went their separate ways to cover more ground, as a way of working collaboratively – but the moment they discovered their prey, it would be up to each individual creature to ensure it consumed its fair share of meat. Just like they did with the other survivors, they would crawl over their bodies, each latching onto a limb – one might have the shoulder, another the knee, or the foot, or the face, or the crotch – and they would demolish their part of the human until they were devoured.

It would take seconds.

And now they were searching – not sure how many people there were, or where they were, or if they were really there, but searching nonetheless. Swarming through, ready for their next meal.

It had been a while since they'd last fed on humans.

And they were hungry for the taste of flesh.

CHAPTER TWENTY-FIVE

FUCK THIS, Paul decided.

He knew what those things could do.

The large, wrathful creature took what it wanted then sunk back beneath the surface – then these smaller, lethal bastards came next. Spreading throughout the boat. Feeding on the leftovers.

So many people on this ship thought that, once they had hidden away and the creature had left, they were safe. They were wrong.

He ran to the cabin and shut the door behind him – but, by the time he'd shut the other door, a few of them had already entered.

They crawled over the window. Over the walls. Over the ceiling. Quick. Agile.

He had to think. Had to stay calm.

If he shot the one on the window it would smash and the rest would come in.

One pounced at his face, and he ducked just in time, spun around and shot it, exploding it into a mess of green and yellow gunk.

Another crawled along the ground, and he shot that one too.

But there was one more.

It had latched onto his leg.

He screamed, shook his leg, battered it against the wall, against the helm, against the wall again, desperately trying to release its pincers.

Its teeth dug into his leg, and he knew it would be seconds until his leg was no longer there.

He had no choice.

He pointed the gun at the creature, trying to miss his leg, and shot.

The creature went limp.

Its teeth, however, still remained in his leg, and he cried out with pain as he pulled each tooth out of his muscle.

Streams of blood ran down his skin.

But he still had his leg, and he was still alive. Miraculously so, but he was. He ripped the sleeves off his t-shirt and wrapped them around the wounds; they may sting, and he'd have a limp, but at least he could still walk – a few more seconds and he wouldn't have been so fortunate.

He crouched by the helm. The creatures crawled over the outside of the cabin, their claws tapping against it, but he knew they had no eyes – so long as he remained silent, he would be okay.

So he stayed as small as he could, not making a sound.

CHAPTER TWENTY-SIX

Cia pulled Boy's arms away from his face, desperate to avoid a poorly timed meltdown.

"We don't have time for this!" she shouted, but he wasn't listening. She tried to pull him to his feet and he resisted. She could think of nothing to do but hit him on the side of his head, which forced his eyes to open in shock. She pulled him up by his arm, refusing to let him resist.

And they ran.

Cia glanced over her shoulder.

What the fuck...

The things approached. Quickly and nimbly. Covering the deck. Too many of them.

She wanted to watch what they did, wanted to learn more about them, just as she had done with Thorals and Masketes, but she was sure that one second's hesitation would lead to a certain death.

She ran backwards and shot at the creatures. The gun kicked back and the empty shells flew away, but she managed to land a bullet in one of creatures, flipping it onto its back.

And she saw it. On the underside of its body. Its mouth, the length of its body, its sharp, pointed teeth exposed.

Bloody hell...

She barged open the door to the nearest floor, Boy following. She didn't have time to shut the door; they were closing in and she had to keep running.

As they ran, she glanced back at them, noticing their claws, the way moved so succinctly, the way their skin wrinkled and curved, and she wondered... where were their eyes?

They didn't have any. Did this mean that if they hid and remained silent, they would be safe?

She thought of all the corpses she'd come across, the marks up and down their bodies, how they were stripped of muscles and organs, and she realised what these creatures could do – how they could each latch onto a limb as their friends took another. She imagined the excruciating pain of being devoured by these things, and the torment of watching yourself being eaten alive.

There was an open door down the corridor.

"In that room!" she told Boy, and she ducked inside, waited for him, then slammed the door closed – but not quickly enough avoid two of them coming in.

She aimed her gun in every direction that she saw them and fired repeatedly.

"Get in the corner of the room," she told Boy as she placed another full magazine in the gun, aware that she didn't have another one. She took the mattress off the bottom bunk, retreated to the corner of the room, and used it to shield them. She aimed the gun from behind the mattress, shooting when she saw movement. The kickback hurt her wrist, but she could tolerate a sprained wrist so long as they were alive.

Boy covered his face and moaned, and she ignored it, just concentrating on where these little bastards were.

She spotted one, dead on the floor. But there was one more. She looked up, down, all around, then jumped as the final creature leapt onto the mattress. She kicked the mattress over and fired through it, again and again, until ammunition ran out and the gun just clicked.

She felt for her pocket, where she had placed the box of ammunition the previous day, and took it out. After a bit of fiddling, she managed to release the magazine and slot a few bullets in.

Once she was ready, she stepped off the mattress, waiting for the final creature to burst out from underneath it.

It didn't.

She kicked the mattress aside. It's upturned corpse lay on the floor.

She crouched beside it, cautiously curious, and looked over its body. It was small, but that didn't matter. So was she. It didn't mean it wasn't lethal.

She poked its mouth with her gun and it instinctively clamped closed. She threw herself backwards. Didn't move. Just stood still. Alert. Staring at the carcass. Terrified that it wasn't dead. But it was. It was just a response from the muscle.

She dropped her head. Wiped sweat from her brow. She was exhausted.

She wished they weren't on this ship, that the stupid Thoral had never attacked, that they had never ended up in that stupid speedboat, and that she'd never ended up unconscious as they drifted out to sea.

How was she ever going to get them back to land?

She had to figure out what to do whilst contending with a large, squid-like beast that could attack at any moment, these tiny creatures from God-knows-where, and a murderous psycho that was somewhere on the ship with them.

Maybe these creatures had killed Paul – or maybe she should make sure before relying on any assumptions.

The sound of the creatures scuttling continued in the corridor, and a few moved over the small window in the door.

Boy was sobbing. She put her arms around him.

"It's okay," she told him, even though she didn't quite believe it herself.

She stared at the door, remaining silent, and waited for the creatures to go, hoping that, once they'd found there was nothing else to feed on, they would disappear back to where they came from.

BEFORE

CHAPTER TWENTY-SEVEN

Must feed them.

Must obey.

Must do what they wish, what they wish indeed must do it all, do it all, do it all.

Paul scuttled like a beetle along the corridor, sniffing out a cell that smelt fresher than the aroma of rotting meat that filled the halls.

He'd become used to it. He'd lived in a cell with it. He'd eaten it.

The meat was not rotting it was edible it kept him alive *sustained me it did oh it did.*

"Hello?"

A voice!

From where?

From down there...

From a cell along the corridor he'd already passed. One he'd already come across, already looked inside, so how did he miss it unless it wasn't real it was in his head *the voices in my head make noises they do oh they do yes they do...*

"Hello? Hello? Is someone there?"

He reversed, stepping back along the corridor, until he came to the cell where a person peered through the window.

A man. Young. Supple. Fresh.

He'd do.

"Oh, thank God!" the man said, rushing to the window. "I'm so hungry, I'm so thirsty, it's been so quiet, thank God you're there!"

Thank God indeed...

"Please, let me out!"

Paul took the chain of keys from his belt where he'd been carrying it like the prison officer used to.

He'd always wanted to gut that prison officer, but something got there first, *got there first oh yes the little scroungers got there first...*

"Do you have any idea what's happening?" the man asked, watching Paul search for the correct key. "I just heard loads of screaming, then suddenly – nothing."

Paul grinned. Did not look up. Did not meet his eye.

He found the key.

Placed it in the lock.

"Oh, thank God for you, thank God!"

He twisted the key and pushed the door open. The man grabbed it and flung it wide, rushing into the corridor. His clothes hung off his bony body. He was pale and malnourished. Not quite as good a find as Paul originally thought.

Still, it wasn't him who was eating it.

"Is there any food?" the man asked.

Paul chuckled.

Yes. You.

He withdrew his gun and shot the man's left shin,

cackling as the bone shifted to the side and pieces of calf muscle flew onto the wall.

The man collapsed, wriggling and moaning, and Paul could not help cackling further.

It was almost too easy.

Paul took hold of the man's good leg and dragged him.

The man tried to wriggle, tried to squirm, tried to hold onto the various doors they passed, but his weakened state and immense pain impaired him, and he could be nothing but be subservient, obey the strange man dragging him out of the corridor, dragging him up the stairs, head bashing against each step. By the time Paul reached the deck, the man was dipping in and out of consciousness, his groggy eyes only flickering open to shut again.

Paul pulled the man across the deck and placed his body on top of the others.

"He-ey!" Paul shouted. "They're here!"

In the distance, the waves grew.

It was coming.

He backed away, returning to the stairs and making it down the corridors.

It did occur to Paul for a fleeting moment – *what will happen if I run out of offerings what will happen what will happen indeed?*

But the thoughts disappeared as soon as he heard more screaming.

A woman screaming.

He was unable to help his grin from forming.

He reached the window of a nearby cell.

"Oh!" a woman cried out, rushing to the window. "Please help me! I shut myself in this cell to get away from them, now I can't get out!"

She was blond. Young. Slim.

Paul had no idea what a woman like this was doing aboard the ship, but he wasn't bothered.

He opened the cell door and entered.

This time, however, he did not let her out. Instead, he shut her in there with him.

After all, he had to eat too, didn't he?

AFTER

CHAPTER TWENTY-EIGHT

The sounds of scuttling stopped, but Cia didn't move. She waited, long after silence had arrived, her arms wrapped around Boy, staring at the window.

What the hell were they?

What was going on?

She'd encountered Thorals many times, she'd fought Wasters, she'd hidden from Masketes – she'd even experienced the rare sight of a Lisker.

But these were something new.

And, as much of a threat as those other creatures were, she preferred a threat she was familiar with; that she'd learnt about.

"Rosy?" Boy said, placing his head on the curve of her shoulder.

"Yeah?"

"What were those things?"

Cia went to answer but found no words coming out.

"I... I don't know."

"Will they be coming back?"

She considered lying to him – but that was something she never did.

"They might," she answered.

"I want to leave the ship," he said. "I want to go back to land."

"So do I."

"How long will it take?"

"I – I don't know. I'm trying."

She felt tears in the corners of her eyes, and wasn't sure why. She dismissed them and refused to let herself cry.

She had to show strength to Boy, and nothing less.

Then she heard something…

A sound, faint but growing louder, coming from the hallway.

Footsteps.

Loud. Deliberate. Stepping so noisily that it had to be intentional; he wanted them to know he was coming.

Cia's chest tightened, along with her arms, her legs, her neck – every part of her bracing for impact; preparing to see the face of their predator.

The footsteps grew heavier, and his smug, dirty face appeared in the window to their room. Cia was terrified, for a moment, that he was about to come in.

But he didn't.

He just paused by the window, looking in at them with his sneering smirk, directing his lecherous glare up and down her body.

She'd never felt more disgusting.

Then he turned to Boy, and enjoyed the sight of him just as much.

"Rosy…"

"I know, Boy. I know."

Cia reached for the gun. Lifted it. Deliberately. So Paul could see it.

Paul just laughed.

"Rosy..." Boy's voice was getting more frantic, more worried.

"I've got the gun," she whispered in his ear. "If he comes in here, I will shoot him. Okay? I will shoot him *dead*."

Boy remained silent. Reassured, but still frightened.

Cia raised her eyebrows at Paul, an inviting gesture, an expression of *what exactly do you want?*

Then the thought hit her with the impact of a punch to the head. She had to kill him. It was the blunt realisation she'd somehow already known but was yet to acknowledge: *it is going to have to be me.*

But how could she kill a man like that?

He wasn't like those she'd killed before.

Her father was a scientist, not a fighter.

She'd left Ryker and his community of sycophants to the monster.

And Dalton had lost his mind in a very different way to Paul – in a way that made him more vulnerable, rather than more cunning.

She'd never had to face someone this unhinged, and this experienced in doling out pain.

Paul lifted a piece of paper so she could see it through the window. He ducked beneath the window, and the piece of paper appeared under the crack in the door.

He reappeared in the window again, waved, and walked away.

She did not get up. Not yet. She waited for the footsteps to go. This could be a trap. He could be waiting for her.

Eventually, after a long moment of silence, she went to get up, but Boy pulled her back.

"It's okay," she said. "I'll be right here."

Reluctantly, he let go, and she picked up the gun and approached the door. She opened it marginally, aimed the gun down the corridor, to the left, then to the right.

He was gone.

She shut the door and picked up the piece of paper.

It read: *Meet me in the cafeteria.*

She scoffed. Meet him? Walk into a trap?

No chance.

Yet the more she thought about it, the more she thought her initial instinct could be wrong.

She had a gun. He probably had a gun. The cafeteria was large and exposed. Plenty of places to run. It could be the safest place to confront him.

What if he'd set something up? A booby trap or something?

It was unlikely. He was crazy and calculating, but if he'd wanted them dead, he'd had plenty of chances already. They were evidently going to give him more pleasure alive.

Cia dreaded to think how.

She looked at Boy. She had to protect him. Which meant she had to face Paul.

There were three monsters hunting them, and this was her opportunity to get rid of one of them.

"Come on," she said to Boy, struggling to keep the gun from sliding around in her sweaty palm. She walked out of the room and, with him following, edged down the corridor.

CHAPTER TWENTY-NINE

A memory, plucked from the hidden part of Cia's mind, presented itself to her at this most inopportune moment. It wasn't a memory she knew she had, yet its image flashed upon her thoughts, accompanied by a few words she still saw through a child's mind.

It was of a man. He was a vicar. Or a priest – her child's eyes weren't too sure of the difference. He had that bit of white around his neck, but was otherwise dressed in black.

Her hand was in her dad's. They were walking with a crowd. The creatures had just risen, and everyone was fleeing, and this man was stood on a bin, an open bible in one hand, holding onto a street sign or something with the other – that part of the memory was hazy. But she remembered he was speaking. And she remembered what he said.

"Though I walk through the Valley of the Shadow of Death, I will fear–"

Then the image cut out.

That was all she had of this memory; those few words and the fleeting glimpse of this man.

She'd never quite comprehended what the Valley of the

Shadow of Death was. Was it Heaven? Was it Hell? Was it both?

Now, however, she understood.

As she held her gun high, aiming at every corner, Boy's steps behind her, she felt as though she was walking through that valley right now.

There was a large, destructive beast in the water.

There were tiny, menacing creatures in there too.

And there was a psychotic, unstable monster on the very ship that was their current sanctuary.

She'd faced all kinds of creatures and all kinds of people. She'd done what she had to in order to survive – but each creature or person had a purpose, however twisted that purpose was.

The creatures needed to feed.

Dalton was angry that Cia had destroyed his home.

Ryker and his messed-up community tried to sacrifice her precious Boy to the monsters because they believed it kept them safe.

Each of them were messed up, nasty barbarians – but they all had their reasons, however much Cia disagreed with them.

So what was Paul's reason?

Chaos?

Anarchy?

Sport?

Was it sexual gratification?

Did he enjoy the hunt, or the torment?

Or was the bastard just bored?

It was far harder to face an opponent when you didn't understand them. Their actions became nonsensical and unpredictable. And that's what powered the fear clutching her mind – the foggy lack of clarity that surrounded her thoughts of Paul.

What does he want?

Cia turned a corner that led her to a set of doors. Behind those doors were the stairs, but she did not enter them yet.

It could be a trap.

He could be waiting on route to the cafeteria to leap out, take her gun, and perform whatever unspeakable acts he was capable of.

She pushed the door open with her foot and pointed the gun to the left, then to the right.

She felt stupid, pointing the gun like this. She was just doing what she saw in James Bond movies her dad used to watch. She didn't even know if her aim would be good enough. She didn't understand how the gun worked beyond 'point and fire.'

"Come on," she whispered to Boy, keeping her steps light as she began her ascent.

She reached the next floor and pointed the gun upwards, searching for movement; someone running on the steps.

Nothing.

She remained focussed as they reached the top, pausing again by the door that led to the deck.

He could be on the other side.

These could be the last few seconds of her life.

As soon as she opened that door, he could squeeze the trigger, blast a bullet through her face, and Boy would be left alone with him.

Oh, God... The thought destroyed her...

Boy would be so scared... He just wouldn't understand...

With a deep breath, she nudged the door open, pointing her gun down the deck of the ship.

The sea was calm. The deck was empty.

She turned to the other side of the deck, and saw no one.

"Stay close," she whispered to Boy, and moved onto the deck, edging along the building.

She paused by every window, checking if anyone was waiting inside.

There wasn't.

She led Boy further along the ship's deck, eventually reaching the door to the cafeteria.

Cia stopped.

She had reached the Valley of Death.

She took another deep breath and nudged the door open.

CHAPTER THIRTY

THE CAFETERIA WAS A GLORIOUS PLACE. Paul hadn't realised quite how glorious, as he'd been otherwise preoccupied with the tasks set out for him.

But oh, what a glorious place it was.

Unfortunately, there was no electricity to power the ovens, and even if there was, the fridges had long since shut off and the meat had gone stale. Shame, really. There was a rack of ribs that looked like it would have been beautiful.

The canned goods, however, were just as scrumptious.

Cold soup may not sound appealing, nor would uncooked chilli con carne, or cold baked beans. But it's all relative, and when one has been forced to eat their rotting cell mate, one relishes such luxuries.

He spread the cans out on the table, opened each of them, then sat there, devouring all he could, not realising quite how hungry he was.

The whole time, he kept his gun in his spare hand, pointed at the door, awaiting her arrival.

After the cold soup and chilli con carne and baked beans, there was fruit. He never thought he'd be grateful for fruit, but that tin of peaches tasted like it had been produced by a god. He followed this with tinned pineapple and tinned apple slices and revelled in the feast he had discovered.

By the time the door began to slowly open, his fingers were sticky and his cheeks were smothered in juices. Most of the cans were empty, but he still had a few left he could enjoy while they spoke.

Her gun entered the room first. Followed by her arms, then her body, and she rotated toward him until the barrel of the gun was in line with his face.

He grinned.

What a silly, little girl.

Paul could tell by the way she failed to lock her wrists and used a high thumb grip that she'd never held a gun before. It was sweet, really. Quite endearing. The pretty little girl was trying to be a grown-up. Trying to act like one of the men.

How precious.

The ratty, lanky oddball followed her in. Paul wasn't quite sure what was wrong with that kid, but he was bloody weird.

Paul didn't say anything at first. He let her stand there, opposite side of the cafeteria, pointing her gun at him while he pointed his at hers.

He lifted his gun, making a point of demonstrating that he could shoot her as quickly as she could shoot him.

With his other hand, he searched for a can that was still full, and found an untouched can of custard. He lifted it to his cracked lips and poured it down his throat as he stared wide-eyed at Cia, his gun still pointed at her, his eyes still fixed on hers. Once the thick, yellow substance was either in his mouth or on his cheeks, he licked out the remains of the can, making sure to keep his eyes locked on the girl's while he flicked his tongue. He could see how grotesque she found it, and he couldn't help but get a little hard.

He slammed the tin on the table, belched, then grinned.

"Welcome," he said, and even he was beginning to notice how erratic the pitch of his voice was becoming. "I've eaten most of the canned goods. It's good shit. Might have something left though."

He searched for a can that still had something in it.

"Aha, here we go!" He lifted it and read the label. "Quality dog food. Meaty chunks in jelly. Beef, I think. Fancy it?"

He raised his eyebrows and stuck out a bottom lip.

Her expression remained deadly. She was not amused.

He turned to the boy.

"What about you, weirdo?" he asked. "Fancy some dog chow? Fancy licking out my can?"

"You don't talk to him," she said, and he could feel the anger in her voice.

God damn, it turned him on.

"Oh, how you don't know what you're doing to me," he lamented. "The feelings I have... Don't tell me you don't feel it too?"

"All I feel is revulsion."

"Revulsion and arousal are pretty much the same thing."

"Maybe to you."

"What? You think you're better than me?"

"Without a doubt."

"All this shit has taught us is how no-one is better than anyone else. Kings, servants, politicians, prisoners – those creatures find all of us just as tasty."

"What do you want?" she said, interrupting him.

"Straight to business. I see. Well, we have things to discuss if we are to continue sharing this ship. Shall we talk?"

Her eyes narrowed.

He could see her thoughts, probably more clearly than she did – she had no intention of sharing this ship with him, but she knew he was stronger. She was trying to be smart.

How adorable.

"I don't like talking with a gun aimed at my head," he said. "Shall we both agree to put our guns down? Then we can talk."

"How am I supposed to trust you?"

"I don't give a shit. Either we put the guns down or we don't, up to you."

She paused. "Fine."

"After three?"

She nodded.

"One. Two. Three."

They dropped their guns to their sides.

"Lovely. Now why don't you come join me?"

CHAPTER THIRTY-ONE

To say Cia was hesitant to drop her gun would be to underestimate the danger she felt simply by entering this man's presence.

The rational part of her mind, however, reminded her that it would be more beneficial to her if they dropped their guns than it would be to him. He could hit his target from where he was sitting. She, however, had no idea how good her aim was. So she dropped the gun on the count of three, relieved to see him do the same.

"Lovely. Now why don't you come join me?"
She turned to Boy. "Wait here."
"No!" he objected.
She sighed. She didn't have time for this.
"I want you to stay by the door," she said.
"I don't want you to go..."
"I'll just be over here," she reassured him. "I promise."
He clung onto her hand, but was forced to drop it as she stepped out of his reach.

Slowly, she approached Paul and his sickening grin. She sat opposite him on the next table over.

"What's wrong with him?" Paul asked, nodding at Boy.

Cia ignored the question. "What do you want?"

"Is he special or what?"

"I said what do you want?"

"Why you sitting there? Don't you want any food?"

"I'd rather not sit that close to you."

"Oh I love it when you play hard to get."

"What do you want?"

"Oh, many things."

"Are you taking this seriously? If not, then I'll–"

"What? You'll walk out? Hide? There ain't many places to hide on a ship I know much better than you do."

Cia remained silent.

"Yes," he said. "Exactly. So you're going to stop being such a prissy little princess, and we are going to talk like grown-ups."

"You're a condescending little–"

"Ah, ah, ah – no name calling. It's not helpful, is it?"

He took another can, put it to his lips, and poured its contents down his throat. Gravy trickled down his cheeks and onto his stained clothes. By the time he put the can down, yellow stains left by the custard had mixed with the brown stains left by the gravy to form a shit-coloured paste.

"How did you survive?" she asked.

"I ate my cellmate."

"I meant the monster. You went in the water. There was no way you could get away. How did you survive it?"

He shrugged. "Because I'm the luckiest son of a bitch ever."

"That's not–"

"I've never thought about that phrase before, you know. *Son of a bitch.* It's not so much offensive to the person, but to their mother. Or their father. Of course, in my situation it's quite accurate – my mother and father were both bitches."

"I know what you're doing."

"What's that?"

"Trying to act all crazy. Trying to freak me out. Trust me, I've faced far scarier monsters than you."

He licked the last remnants of gravy out of the can then paused, excited by her tenacity.

"You've got all kinds of attitude problems, ain't you?" he said.

She rolled her eyes. "This is going nowhere. Did you actually want to talk?"

"Another prime example."

She huffed.

"Look," he said. "You can think what you want of me. I'm a good person. And I'm prepared to talk."

"You're a good person?"

He nodded, licking the last few bits of the can and discarding it.

"No you're not," she said. "I've seen your file."

"Hah!" he snorted. "Busted."

"So what do you want? You want to take one part of the ship, I take the other? You want to split it up that way? Since we're both stuck here together, and neither of us has any idea how to get back to land?"

He didn't reply.

"Well?" she prompted.

His grin spread even further across his cheeks.

"I'm waiting," she said.

He sat back, smirked, and watched her, biting his bottom lip, rocking, with a cocky, knowing look on his face.

"I have no intention of sharing this ship with you," he said.

"Excuse me?"

"You heard. Truth is, I intend to overpower you, torture you, humiliate you, fuck you, then give up what's left of you to the beast."

Cia looked into his eyes. Wide. Bloodshot. Arrogant.

He meant every word, and Cia felt foolish.

Paul had never intended to negotiate, or to discuss their sharing the ship, or to figure out what to do next in terms of getting to land. He was quite happy for the ship to drift wherever it drifted, so long as he had his playtime.

"So," said Paul, standing. "Shall we dance?"

She had to get out of there.

She had to get her and Boy to safety.

But where, exactly, was *safety*?

As she stared back into this man's grimy face, with remnants of canned food spread across his cheeks, her thoughts presented the answer in a moment that sent dread trickling down her spine and shook her legs: *nowhere*.

There was nowhere safe from him or any of the creatures.

They were stuck on this ship, simply waiting to be killed.

She collected her gun and went to get up, ready to get her and Boy away from this man – but she had barely stood in the time it took for him to climb onto the table, leap to hers, land on top of her, and take her to the floor, sending her gun across the floor and out of sight.

He pinned her down. She tried to kick and punch, but he was too strong.

She was helpless – just the way he wanted her.

CHAPTER THIRTY-TWO

Oh, it had been too long, and it felt good to be on top of a girl again.

Yes, this was a mixed-race girl, and she seemed to have some convoluted fucked-up relationship with some strange boy – who would be next – but it had been a while since he'd found a living woman on the boat.

Honestly, after a while, corpses just don't have the same kind of juices as someone alive.

He felt saliva trickling through his open-mouthed grin and, instead of licking it away, he allowed it to fall upon his prey.

She squirmed and kicked and thrashed and screamed, oh boy how she screamed, but each movement of retaliation was another punch of arousal. She didn't seem to realise that, with each bit of struggle, his dick got that little bit harder.

And the harder he was, the harder it would hurt.

"Stop it, please, you don't need to do this!" she shouted, and he was a little disappointed. This girl was undoubtedly damaged by everything she'd experienced. The world had torn her up and ripped her apart, and he enjoyed the prospect of adding more trauma to her misery – but he didn't think she'd be so foolish to think she could rationalise with him.

Then again, the fact that she had both come here and acquiesced to lowering her gun only showed her willingness to believe that there was still a shred of good in this world.

As if she needed to cling onto it.

Maybe that was why she kept the idiot with her. Maybe he was her link to hope.

Well, Paul was relishing the moment he would see that last bit of hope fade from her eyes, that moment when she felt the knife pierce her throat, the moment that would follow him turning her over and teaching her what it truly means to be sodomised by Hell.

She fought with all she had, though. It was cute. Her fists were flying upwards, and they even landed on his chin and his cheek, and hell, some of them even hurt a little. But he was on top of her, his fingers digging into her shoulders, his strength overpowering her, and her resistance had little effect. She was a tiny little girl, and he was a strong man. Of course he was going to win.

"You know," he said, "the more you struggle the more it's going to hurt."

She paused. Stopped struggling. Like she'd listened to him.

Huh. How peculiar...

Then she grabbed the back of his head, pulled his face to hers, and pressed her lips against his.

This was unexpected.

Her tongue slithered into his mouth, meeting his, and he wasn't sure how he felt about this...

He preferred it when they weren't compliant.

Before he'd thought any more about it, something hard knocked against his head and sent him onto his back.

The ceiling spun. He wiped his eyes. What was going on?

It took him a while to realise he'd been hit by something, and that the girl was running away.

CHAPTER THIRTY-THREE

As soon as they made it out of the cafeteria, Cia rushed to the railings and spewed over the side of the ship.

She turned to Boy, took his hand, and said, "Good job."

He'd done so well.

When she'd seen him approaching with the chair, she knew Paul would notice him. So she'd had to do the only thing she could think of to distract him...

And thinking about it made her almost throw up again...

She glanced into the cafeteria. Paul was rubbing his eyes. He was coming around, and they needed to make the most of their head start.

"Come on," she said, grabbing Boy's arm and rushing to the stairs, back down a few floors, into another corridor.

Was this it?

Was this all she could do?

Just keep running through corridors, finding a hiding place, hoping he wouldn't find them?

Because he would.

Stop it.

Can think about that later.

First, just...

Just what?

Hide?

She hadn't even got a gun anymore. She dare not risk going to the armoury again. She needed to be somewhere hidden.

She chose a room at random, opened it, and searched for a place they could hide.

She didn't think of locking the door. She didn't have time to search for keys, and having the door unlocked would hopefully mislead him – perhaps if he was to discover an unlocked door, he would assume they were not behind it.

She hoped.

"Under the bed," she instructed Boy. She looked over her shoulder as Boy shimmied underneath, expecting Paul to appear, expecting him to walk through that door and subject her to whatever torture he'd been fantasising about.

Once Boy was under, she shimmied under too, then reached her hand out and felt for the duvet. She pulled it down until it overhung the bed and covered their hiding place.

She wrapped her arms around Boy, hugging him whilst placing a hand over his mouth.

"We have to be quiet," she told him.

"Rosy..." Boy whispered. "I'm hungry..."

His belly rumbled, and she realised that she was hungry too. She couldn't remember when they last ate. They wouldn't survive without food or water.

She just needed time to think.

Time for Paul to search and not find them.

Once he didn't find them during his first search, would he give up, or would he search more thoroughly?

Would he then start looking under duvets and under beds?

She shook her head. Tightened her grip on Boy.

How had it come to this?

They'd found a way to survive. They hid in buildings and dodged the monsters. They ran when they heard a creature's growl. They avoided any other survivors, though they hadn't found any for a long time – until they found this ship, that was.

She'd felt as safe as she could in this world.

And now...

Now she was hiding because she wasn't strong enough to fight.

She'd fought Dalton, she'd fought Ryker, she'd fought monsters... So why was she hiding?

She'd beaten worse opponents than this, dammit!

As much as she repeated this to herself, it did not make her feel any more confident. It did not make her believe she should leave her hiding space and confront him.

Maybe it wasn't about how formidable the opponent was, but what the stakes were should she lose...

This man would do things. Not just to her, but to Boy. She feared death because of what it meant would happen to Boy.

"I'm hungry..."

"I know, we'll get you some food in a bit, just be quiet for now."

"But my tummy hurts..."

"I know, but please, just be quiet."

So what next...

How were they going to get food... Water... Or even survive...

She knew they couldn't stay here.

She knew she had to think of something better.

But, other than finding a way to turn back time and not let Paul out of his cell, she couldn't think of anything.

For now, she supposed, she just had to concentrate on surviving the night. And, as she heard the creature pass the ship, feeling the tremble of the walls around her, she realised just what an impossible task that might be.

CHAPTER THIRTY-FOUR

Once again, she waited until Boy was asleep to leave the room. She didn't want to leave him in such a vulnerable state, unable to call out for her should Paul find him – but it was the only way to leave without any resistance. If he was awake, he would be distraught and she would feel guilty. It was just easier this way.

She needed food and a weapon. They stood no chance without it.

She lifted the duvet from its position concealing them under the bed and peered out.

The room was now dark. The small window in the wall let very little light in during the day as it was, but now it was almost black. It was the perfect cover for her to sneak out.

Then again, it was the perfect cover for him to be waiting for her.

Stop it. I have to do this if we're to survive.

She scoffed at the thought. She'd never had less belief in the prospect of their survival.

Even when Dalton was losing his mind, when Ryker had taken Boy, or when the Wasters had her hostage, she still had hope. She had always felt in her gut that there was a chance she could win. But in all of those fights, she'd been on land. There had been somewhere to run. Somewhere to hide. Now, she was on a ship Paul knew better than her, just waiting for him to find her, or for the monster to attack, or for those smaller creatures to come aboard.

As she reached the door, she told herself to calm down. Such thoughts weren't helpful. They were destructive. She simply had to try.

When she was a child, she ran the 500 metres on Sports Day, and she was scared about losing – her father had told her, "Don't try to win, just try to do your best. That means you'll never lose."

He was right.

Then again, he abandoned her and left her to die, so maybe his words weren't so wise.

She placed her palm against the door, slowly nudging it open, only slightly, then pausing.

Listening.

Feeling the soft sway of the ship, hearing the distant movement of the ocean, the eerie silence of the corridor.

She pushed the door open a little more, just enough for her to slither through, and as she did, she paused. Checked to the left, then to the right.

The corridor was dark, but the shadows were still. It seemed she was alone.

But could she ever be sure?

Stop being paranoid...

She crept out of the room, keeping her back against the wall, ensuring nothing could surprise her from behind. She moved slowly, stepping lightly, keeping her breathing quiet.

Bodies were visible through every door she passed.

Nothing else. Just bodies, their flesh and muscle torn away from a previous attack by those creatures.

The sight of it was enough to renew her terror.

She understood Thorals, Masketes, Liskers, Wasters... She'd lived with them long enough...

She didn't understand these other monsters.

And one always fears what they don't understand.

Eventually, she reached a stockroom, twisted the handle, and nudged it open. It had been ransacked, but not much had been taken, which didn't surprise Cia, as it was mostly plates and dishes – but the stockroom was a mess as a result.

She searched, feeling objects, the outline of boxes faint in the darkness.

Eventually, she found what she was looking for.

Carving knives.

She took one out, ran her finger gently against the sharp blade, felt its sting, and was satisfied she had a sharp enough weapon.

Then she heard the distant creak of a door.

She turned, wide-eyed, catching her breath.

She heard footsteps down the corridor. Coming closer.

She shut herself inside the stockroom, put her hand over her mouth to stifle her own breath, and gripped the handle of the knife – just in case.

The steps grew louder.

They were shambolic. Chaotic. The walk of a madman.

"Girly girly girly..." his voice sang out with no particular melody. "Where are you..."

A door swung open; one close to her. Maybe a few rooms down.

"You must be somewhere... I've been looking for you all over..."

Another door. This time, even closer. Swinging open and colliding with the wall.

"Come on! You know this isn't going to last..."

A pause, then another door hitting a wall.

He must be close.

"I promise it won't hurt if you don't struggle..."

The door in the room next to hers collided with the wall.

She rushed to the corner of the stockroom just as he kicked the door, and it swung open, trapping her behind it.

The darkness gave her no indication of shadows; she could not track his movement. But she could hear his feet enter and pause, and she could smell the rank aroma of his body odour.

She waited for him to leave, but he didn't.

He stayed for longer in this room.

She held her breath.

Does he know I'm here?

He sniffed.

Like he was a predator in the wild, hunting its prey.

"Hmm..." he said.

Cia lifted her knife, ready to strike should she need to.

But she didn't. Paul left the stockroom and continued down the corridor, kicking open the next door. And the next. And the next.

It suddenly occurred to her – what about Boy?

She couldn't let Paul find him.

She peered around the door, only moving it marginally, squinting through the darkness.

Paul had disappeared around the corner, but she could hear him kicking more doors open.

She crept along the wall, staying out of sight, tuned in to the noise, waiting for Boy's scream.

But it didn't come. Paul passed Boy's room, kicking the door open, and Boy remained silent.

She couldn't have been prouder.

Eventually, the door at the end of the corridor opened and shut.

Cia rushed down the corridor with caution, ensuring he hadn't pretended to leave the corridor, and entered the room where Boy hid.

"Boy?" she whispered.

He lifted the duvet. Tears ran down his face.

She wanted to rush up to him, to hug him, to reassure him. Instead, she was gripped by the prospect of opportunity.

She knew where Paul was. That gave her an advantage.

She had a knife.

Could she sneak up on him? Slit his throat from behind? Stick the knife in his neck?

She had to try. It was her best chance. They couldn't hide forever – he'd see her when they went for food.

"Stay hidden," she told Boy.

"No!"

She shushed him. "Just trust me. Please."

She gave him a forced smile and shut the door before he could object further. With the knife held out before her, she sneaked along the corridor and to the stairs.

CHAPTER THIRTY-FIVE

Cia emerged onto the deck, holding the door ajar with as small a gap as she could, peering out.

She watched Paul walk along the deck, his back to her, swaying, stumbling from one side to the other.

Once she left the stairs and emerged onto the deck, she was exposed. It was a risk. The only way she would succeed was to maintain the element of surprise, and that made it far harder.

She looked for a hiding place, somewhere she could run to.

The doorway to the cafeteria was in a slight alcove. If she could get to it, she could use it as her next hiding place.

Keeping her eyes fixed on Paul, she waited for the opportune moment.

Was this a ruse? Did he know she was there? Was he trying to make her reckless?

Those thoughts weren't helping, so she banished them. Scolded herself for undermining her own confidence at a moment when she needed it the most.

But what if she didn't make it back, and Boy was left alone with this man?

She shook her head.

Just shut up.

It was always the thoughts she couldn't control that shouted the loudest. But this time she refused, and she drowned them out with rational thinking – *I just need to take it step by step.*

And the next step was to get to that doorway.

Without further hesitation, she left the corridor, shutting the door silently behind her, and crept along the wall. A few seconds felt like hours, but she made it to the door, and she pressed herself against it, hiding herself behind the wall.

Paul stopped.

Looked over his shoulder.

She panicked, her body seizing with terror.

Then he turned back around and walked on.

She needed to be cooler than this. Calmer. She needed to be in control.

But there was this feeling... Deep in the pit of her stomach... A feeling like this was a bad idea. Like there was more to what Paul was doing than it seemed. Like something horrendous was about to happen.

Once again, she willed the thoughts away.

Paul disappeared behind the building, so Cia moved from her hiding place and crept along the wall until she reached the edge of the building, and stayed hidden behind it.

Paul walked to the end of the ship and stopped.

His body hunched over. His neck twitched. Irate tics making him convulse.

She hadn't seen him like this. Something was making him anxious. What was it?

He knelt on one knee and bowed his head.

In the distance, the ocean rumbled. The reflection of the moonlight in the sea flickered as the waves grew rougher.

Cia knew what this was. This was the creature coming.

She turned to leave, then didn't – because Paul wasn't leaving. Why hadn't he left? Why wasn't he running?

A bump in the water grew closer, the thrash of the waves rising in a crescendo.

Fear gripped her, yet she still didn't run.

She'd run when he did.

From the water, it emerged. Grand and exquisitely terrifying, its large tentacles waving in each direction, its domed head turning toward Paul.

She waited for its mouth to wrap around his body, for his head to be ripped off or for him to flee in terror.

But he didn't.

She'd seen it before. At the last place she'd called home. The way they worshipped, offered sacrifices, formed allegiances...

But Paul was insane. He wasn't calculated and intelligent, he wasn't capable of rationalising with himself, never mind a creature.

But, as the creature hovered by the end of the boat, it did not hurt Paul.

And Paul remained on one knee, his head bowed toward it, waiting to be allowed to speak.

CHAPTER THIRTY-SIX

Survival.

That was what this was.

Survival, pure and simple.

Yes, he had attempted to indulge himself in the enjoyment of more pussy – but that was a simple advantage to his predicament. A finder's fee, if you will.

And now he wished he hadn't been so foolish.

The attempt to use the girl for his own temporary satisfaction had let her get away. He should have stuck a knife in her leg and dragged her onto the deck. This was what he'd been kept alive to do, after all.

"I'm sorry," he said.

As soon as the first syllable of his apology left his mouth, the creature's tentacle thrashed the water's surface, sending a wave of cold water over the ship's frontend and soaking him.

He did not move from his subservient position. He didn't dare. To move would be to disrespect the one who had granted him life where others weren't so fortunate.

"I'm trying," he said.

Another thrash, another demonstration of its fury. He

waited for it to kill him, to eat him alive as a show of its wrath, but it continued to spare him – though he realised its mercy was fragile.

"She's not like all the others I brought you. They were barely alive, desperate to trust me. She... she isn't."

Another thrash, and this time the tentacle knocked the ship, which leant over before rebalancing. He grabbed the railings, determined to stay on his knee, endeavouring to show his dedication to slavery.

"Please, she's worth it. So's he. They are young. They aren't dead or dying like the others. These two are fresh and delicious. You'll enjoy them."

Its face lowered to his, and he knew why it was angry.

I tried to enjoy them too...

"It won't happen again," he promised. "I was arrogant, I know. I won't let myself fall into my own trap. It's just... She was so..."

He dropped his head. Sighed.

"I mean, to be fair, I did bring them to the helm already. I had to stall, pretending I knew how to work this ship, while I waited for you. They were on the deck, they were right there, and you–"

A tentacle landed on top of Paul, wrapped around his body, and squeezed.

At first, breathing was difficult. Then it became impossible. He wheezed, desperate for air. Just one lift of the tentacle and his head would be in the creature's mouth.

"I'm sorry! I'm sorry!"

It loosened its grip, allowed him to breathe, but did not release him.

"I will get them to you by the end of the night."

It went to squeeze again.

"I know, I know, I already said I would! But please... If I

don't have them to you by the time morning's here, then you can have me, you can destroy the ship, you can do whatever... Just please, give me until dawn. They won't be hard to get."

It paused. Like it was deliberating. Like it was considering his offer.

"I will smoke them out," he said. "Trust me, they will have no choice."

The tentacle squeezed again, and he was worried that he hadn't convinced it.

Then the suckers released his body, and the tentacle slithered away, leaving his wet body sticky and hot.

"Thank you," he said. "Thank you."

The creature went to leave.

Then it paused.

It had noticed something.

Paul looked over his shoulder, and that's when he noticed her too.

CHAPTER THIRTY-SEVEN

Cia finally understood.

She'd wondered why the creature had left the boat intact...

It could have destroyed the ship, but all it did was knock it to the side. It was intimidation tactics, not war. It was waiting for her to emerge.

After all, why destroy this ship when it had been its best source of food?

And she finally understood how Paul had survived...

He was a demented chef operating an all-you-can-eat buffet.

All the other survivors... All those he found... He'd presented them. Lead them into a trap, or beat them into submission and dragged them out, or thrown the remains of their sodomised corpse onto the deck.

He'd been allowed to live because of his offerings. Just as Ryker and that demented community had kept the creatures away through sacrifices, he had survived by supplying the monster with reasons to allow him to live.

But now he was desperate. There were no other survivors aboard. She and Boy were the only fresh meat left.

And they had until morning before the creature lost patience with him.

That was why he'd stalled at the helm...

That was why he hadn't killed her yet...

And that was why he was hunting her...

He needed her.

She fell to her feet as the creature knocked the ship and wrapped its tentacle around Paul and squeezed – then released him.

It released him.

It *let him live*.

Because he was offering something better. Something that didn't stink, didn't look so repulsive, didn't create a feeling of nausea.

And, just as the creature released Paul, and its tentacle slithered back into the sea, and it turned to go, it looked in her direction.

As did Paul.

She froze. Scared to move in case they would chase her, but scared to stay still in case they would catch her.

Then she heard it. The scuttling, the taps of pincers against metal, the mass of tiny creatures beginning their approach up the ship.

All three predators had her in their sights, and they knew where she was, and they were ready to devour her.

She wondered which would be the better way to die: eaten by the monster, raped and murdered by the deranged lunatic, or to have her flesh and muscle torn apart by the creatures with no eyes and sharp teeth.

"Boy..."

She had to get back to him.

Had to protect him.

But for what? And for how long?

How was she even meant to survive the night?

No time for thinking. *Just run.*

She turned and sprinted back along the deck, and heard Pauls stomps behind her.

She returned to the stairs and leapt down them two, sometimes three at a time, ignoring the realisation that there was no way she could beat all these monsters at once.

Maybe this time this world had the better of them.

Maybe this time... they were going to lose.

BEFORE

CHAPTER THIRTY-EIGHT

"Please, no, please spare me..."

The sun was high in the sky behind the giant squid's head, but Paul was forced to shiver in the beast's shadow.

The corpse he'd dragged up the stairs lay on the deck beside him, rejected by the creature. The body was old and decaying, and it was an insult to even entertain the notion of presenting it as his offering.

"I know I haven't brought you much fresh food, but I will, really, I will..."

This thing couldn't growl, but it could boom, a low-pitched boom that would send all creatures in the ocean fleeing, and it used such a sound to voice its displeasure.

Paul was ashamed.

Not because he was failing the creature, but because he was desperate not to die. His ego made him unwilling, but his survival instinct overrode his pride.

"Haven't I done well? I've brought you loads of people... I've collected the other survivors, and one by one I've led them to you... But I'm running out."

Its head lowered itself toward Paul.

Excuses, excuses, excuses.

He'd sworn. He'd been granted liberty to roam about the ship unharmed so long as he brought food from below and presented it to this creature.

But a single human must be such a puny meal to such a large beast. He'd been required to bring more and more, to seek out what few survivors remained.

Eventually, those survivors had run out, and he'd been forced to bring the creature a corpse. The creature didn't even entertain the idea of eating something that was so stale.

And now?

Now came the moment he'd feared.

The moment no more survivors could be found.

When he was no longer of use to the creature.

"I will find more," he said, knowing he couldn't. But whether he could follow through on such a promise was irrelevant; it was something the future Paul would be concerned about. Right now, such a promise was all that ensured his survival.

"Please, I will."

The creature hovered. Waiting.

"Now?"

It still waited.

"Fine. Okay. I'll... I'll look."

He backed up, retreating to the stairs.

He could just hide. He could ignore what he'd sworn to this thing, and he could find a small corner where he couldn't be found.

But it would just destroy the ship, and he would be stranded in the ocean with nothing to keep his safe.

The only way was to obey.

It was the only reason this thing didn't just swipe its

tentacle down the middle of the ship and collapse it into the depths of the sea.

Paul rushed through the corridors, frantically searching each room, looking for someone, something.

There were plenty of bodies...

To bring the creature a body that had already been devoured by its smaller minions... The impudence may not be rewarded.

Down another floor, past every window, looking, searching, desperate.

Nothing.

Down another room, along the corridor, in every room.

Still nothing.

What was he going to do?

He had nothing to present.

Was this how he was going to die?

After all those times he'd watched the creature relieve its victims of their last breath, he was now going to witness it done to himself.

He grew suddenly angry at his own acquiescence to the creature's demands.

Fuck him. He wants a fight, he can have a fight.

The thought left quickly as he realised there was no fight to be had.

Down another floor. Into another corridor.

A sound.

Distant. Faint.

He opened a room.

His heart surged. He was safe.

A child was sobbing. Curled up into a ball in the corner of the room. The boy lifted his head from his arms, and his terror mixed with relief upon the sight of another human being.

"Are you here to save me?" the child asked.

Paul considered how to do this. Whether to bring him kicking or screaming, or led by the hand.

"Yes," he replied, deciding on the latter. "I am."

Paul held his hand out.

"Come on, it's time to go. There's a boat waiting."

"Are my mum and dad okay?"

"Yeah, they are waiting on the boat. Come on."

The boy placed his hand in Paul's and wiped his eyes with the other.

Just as the child was about to leave, he said, "Wait!" and grabbed a teddy bear from beside the bed, which he dragged with his spare hand.

Paul guided the child along the corridor and to the stairs.

"Where is everyone?" the child asked.

"They've already evacuated. They are waiting, with your mum."

"Is Mum okay?"

"Barely touched. Desperate to see you."

He nodded, and his sobs stopped.

Paul led him to the deck and guided him to the end of the boat.

"Where are they?" the boy asked. "I can't see them?"

The creature rose from the beneath the sea's surface, water spraying off its head and onto the ship.

The child went to run, but Paul grabbed him by the throat. He threw the child to the floor, and the creature wrapped a tentacle around him, lifted him to the air, leering at his thrashing body, and placed him into his mouth.

It swallowed the child within seconds. This wasn't a good meal; it was small, and reeked of desperation.

Paul could tell this by the look in the creature's eyes.

"I'll do better," he said.

The creature did not retreat.

"I promise. Please, just accept this for now, and I will find something else. Something fresh. I will."

Still, the creature did not back down.

"I found you something. Please, just... Just let me keep looking. I'll have something in the next few days."

Finally, the creature submerged itself, and disappeared from the ship.

Paul let his breath out, not realising he'd been holding it.

He searched the ship again.

And again, and again.

There was nothing. No one left alive. Nothing at all.

Night came and went, as did the morning, as did the afternoon, and still, nothing.

Then he saw it.

In the distance, approaching the ship. A motorboat. And there were survivors on it.

Two of them.

A male and a female.

Young.

He collapsed to his knees. He was saved. This was his salvation. A perfect offering, one the creature surely couldn't help but appreciate.

But these weren't like the other survivors on this ship; these prisoners and stowaways hadn't faced the true horrors of this new world. Neither had Paul, being honest. He'd seen the creatures when they were evacuated, but that was a long time ago. But if these two had the ability to survive such a perilous world, then he may need to be smarter. To rush into this would be reckless.

He had to make them trust him.

Then he had to trick them.

And so he returned to his cell, next to the corpse he'd been forced to eat.

He'd be honest with them. It would shock them, but they would question how he'd survived so long otherwise.

And he'd pretend he was unable to unlock the cell.

He would tell them he could get them back to land.

He would get them to trust him because they hadn't a choice.

Then he would offer them to the creature, just when they thought they were in a place of safety.

It was perfect.

These two were going to grant him a few more weeks of life, and he would savour each moment of it.

AFTER

CHAPTER THIRTY-NINE

She'd seen everything.

The little bitch.

Paul was sick of her. She should be dead already. How difficult was it to control two little inbred fucks?

He growled. Marched across the deck, fists clenched so tightly his nails hurt his palm.

She rushed through the door to the stairs.

Paul kicked it open, slamming it against the outer wall then charging down the steps.

He was beyond furious.

He was enraged.

And he could feel his maddened state sinking lower and lower as the wrath took over.

It was a familiar feeling.

It was the one he had when he killed his wife. When he strangled his son. When he mutilated their bodies, churning and grinding them up until they were even more pathetic than they were in life.

It was a rage that got things done.

It was a rage that he didn't fight, that he allowed to take over – one that powered every step.

He was going to make this girl's final moments hell.

Just presenting her to the creature wasn't enough. He was going to beat her, force her to walk naked to her death, and laugh as her humiliated face begs the creature to spare her body, grinning as it crushes every bone in its jaw.

But he wouldn't do this until he'd made her watch the boy suffer first.

Down the steps she went, into another corridor, and Paul kicked the door open.

"I will find you!" he roared, his voice echoing off the walls, creating a symphony of fury.

She scurried away, and he quickened his pace into a run.

"I will get you!"

She reached a room. Opened the door.

"You dumb fuck!"

Just as she went to shut the door, he stuck his arm in the way, making her unable to shut him out.

"I have keys... I have ways... You can't run..."

She swung the door back and struck it against his arm.

It hurt, but not enough for him to retract.

She swung it again.

He reached further in and the door struck his head. In a moment of dizziness, she pushed his arm out and slammed the door shut, locking it.

Oh, how his rage grew. Intensified. Morphed into an anger even his insanity couldn't manage.

He charged at the door, his words indecipherable. He headbutted the window, again and again, sure he was feeling it give but seeing no crack.

Inside, the girl stood against the far wall with the wretched, stunted boy.

She held a knife out.

He couldn't help but laugh. A laugh that turned into a chuckle that turned into a cackle that turned into hysteria.

"A knife?" he said.

He shook his head.

"She thinks a knife will do it..."

He blurted out a sharp *Hah!*

"A knife a knife a knife oh a knife what a knife... Silly little girl likes to play with knives..."

He punched the window, again and again, determined to find a way in.

It did not give. It was reinforced. His knuckles were not enough.

He focussed his eyes on hers, pressed his forehead against the window, and articulated carefully and succinctly, "You *will* leave this room."

She was cowering. He could see it. She was afraid.

She should be.

She'd read his file, but that was barely a touch on what he could do.

"You *will* return to the deck by the time the night is through."

She clung to the boy, gripping him, pulling his body close to hers.

They looked ridiculous. This tiny mixed-race girl comforting some lanky prick that towered over her. This boy had all kinds of things wrong with him. If it was up to Paul, the kid would be the first to die.

"You *will*. Trust me."

He backed away.

If she wouldn't leave the room of her own accord, then he would force her.

He would smoke her out.

With a wink in her direction, he backed away, off to create some fire.

CHAPTER FORTY

Cia knew Paul meant what he said.

He was many things. A psycho; a deluded freak; a maniac... but not a liar.

He needed them at the deck before morning. It was his agreement with the creature – it was his only way to save himself.

So they had to endure whatever he did until then.

But the look on his face... The wink... The determination in what he said...

He was going to find a way to get them out of that room. Should she pre-empt it? Find somewhere else?

But where else was there to go on this ship but another room where he would just find them again?

She turned to Boy. Placed her hands on his arms. They were shaking. His face contorted as he resisted his tears.

She knew how he felt.

"Listen," she said. "We just have to survive until morning. Until then. Then Paul will be gone. That's all. I..."

Over Boy's shoulder, Cia saw something.

Something she hadn't seen in days.

Something that changed everything.

"No..."

Why now? Why did she have to see this now? Such an impossible sight, one that she'd been desperate for, at the most inopportune moment...

Boy looked over his shoulder, and he saw it too, through the porthole.

Land.

And it wasn't just some island, either – there were buildings. It was a country. The ship had somehow drifted toward it.

Only, fairly soon, it would be drifting past it...

Cia turned away, sat on the bed, rubbed her head with her thumbs, tried to massage her head into being able to think.

Could she wait until morning?

No. There was no chance that, should the ship remain on its current trajectory, it would remain in sight of the land. Especially if the creature appeared and started thrashing about, sending the ship back to sea. They would undoubtedly pass this land by the time the night was through. Most likely in the next hour, if she thought about it logically – the land was to the ship's left, which meant that if the ship kept drifting forward as it was, it would drift away from the land and back to sea.

She covered her head with her arms.

This couldn't be happening.

Why now? Why after all this desperation to return to land, this yearning to leave this ship, is it now that the prospect of relative safety appears?

She looked at Boy.

Tried to think of another way.

They could wait until they found land again.

But how long would that be? With the large creature and the little creatures, would they even survive that long?

No. They had no choice. This was their only opportunity.

"Dammit..." she muttered.

She had to get to the helm and turn the ship around, steer it toward the land, then just stay alive until they reached it.

But to do that, she had to get to the deck.

She ran her hands through her hair. Shook her head. She couldn't believe this.

They were going to have to leave this room.

They were going to have to go exactly where Paul and the creature wanted them to.

CHAPTER FORTY-ONE

Over the deck and through the cafeteria, every step a stomp, every opening of a door a punch.

Paul found what used to be the cooking area. A large oven and a stove fixed to the wall. A few microwaves on the floor, smashed and dented.

He searched the draws, emptying them of vast amounts of cutlery, even bypassing the knives. They weren't what he was looking for.

Then he did find what he was looking for.

A box of matches.

A grin spread across his distorted grimace. Satisfaction simmered beneath his boiling rage. He was one step closer.

He found a bin bag on the floor and emptied it, using it to store anything he could find that was flammable.

There were cooking books. Napkins. Hand sanitiser. In the cupboards were kitchen polish and other aerosol products. Rotting garlic. Flour. Cooking oil. In the fridge was mouldy bacon. Half-used cartons of milk with bits of white floating at the surface. Crusty peppers. He stuffed them all in the bag.

Then he searched the supply cupboards. The ovens were gas, and there had to be something that powered them...

And there it was. Propane. Stored as liquid in small, pressurised tanks.

"It's so beautiful..."

He took empty bottles from the recycling and filled them all. He stuffed as many as he could in his binbag, took a tank under his arm, and turned to leave. Then stopped.

The cafeteria was swarming with the smaller creatures.

"Oh, fuck..."

How was he meant to get those two kids out if the creature allowed its minions to hunt him too?

Oh, wait! How could he so silly? He had the most powerful weapon a man could yield in his hands...

Neanderthals had it right, really. They didn't need fancy talking or books – they made fire, hunted their prey, then clobbered any woman they took a fancy to and dragged them into their cave.

Yeah, the cavemen had it right...

He took a match, struck it, and set a piece of napkin alight. He opened the door and, just before the creatures pounced, he threw the contents of a bottle of propane over them, followed by the lit napkin.

A few burst into flames. The others flinched away.

He lit a match, held out an aerosol can, and placed the flame in front of the can's nozzle.

"Back the fuck up," he said.

The creatures didn't back off. And, once the napkin had reduced to ash, they scuttled toward him again.

He sprayed the aerosol can before the match, sending flames surging toward the creatures. They retreated, the survivors running around in a panic as the others burnt to death.

"Yeah, that's it..."

Now they backed away from him, and he couldn't help but feel a bit of his ego return. He felt a little less pathetic, knowing he could scare such ravenous pests.

"Relax," he said, leaving the cafeteria. "I'll get you your food..."

CHAPTER FORTY-TWO

Cia clutched her knife as she looked to one end of the corridor, then the other.

No sign of him.

They were going to have to leave the room at some point, and now was as good a time as any other.

She looked back at Boy. Sighed.

"Oh, Boy…"

His fingers were fidgeting. His lip trembling. His body shaking. His eyes, despite becoming adult's eyes, still screamed vulnerability. And, as she looked at him, the person she was yet again risking her life to protect, she did not feel annoyance, nor hesitation, nor anger for having to go through all of this.

All she felt was love.

Overwhelming, brave love, firing through her body. A caring she couldn't articulate. A need to nurture in any way she could.

She did not regret the day she met him, all those years ago, in a petrol station while Masketes ate his parents.

If she had to endure this apocalypse to find him, then she'd endure it all over again.

He seemed to see the caring in her eyes, as he placed his arms around her.

"Stay with me, okay?" she said.

He stepped back. Nodded.

He was being strong.

For her.

Anyone else, he would break down and refuse to go.

But he trusted her. And she would not betray that trust.

"On the count of three, we run."

He nodded.

"One."

She flexed her fingers around the handle of her knife.

"Two."

She gripped the door handle, ready to pull it open.

"Three."

She swung open the door.

The corridor was full of the small creatures. She instinctively grabbed Boy's shirt, pulled him back and slammed the door.

It took her by surprise. She wasn't expecting it. And three of those creatures had managed to get in.

The first was crawling along the floor toward her. She backed up, watching it, trying not to think of how sharp all those teeth were, or what the bodies they'd already devoured looked like.

It leapt toward her and she swung the knife, miraculously landing it into the creature's body just before it was able to land on her chest.

Its carcass slid off her knife.

The second leapt onto Boy, digging its claws into his leg.

Cia swung the knife into its back, being careful not to go

through to Boy's leg. Its muscles relaxed just before its teeth sunk into his muscle.

She looked at him, both of them knowing that she'd only just saved him from losing a limb.

But there was a third.

Where was it?

She looked around, under the bed, under the desk, under the chair – but it wasn't until she looked up that she saw it. It dropped onto her face, forcing her to drop the knife.

A set of teeth dug into the side of her cheek, and she howled in pain, and she thought – *is this the end?*

But before it could dig its teeth in further, a knife sunk through it, its tip piercing her skin, and the creature relaxed and dropped off.

She fell to her knees, grateful to Boy for helping her, but unable to articulate it through the pain.

She felt her cheek, which left enough blood on her hand to fill a small cup.

Boy began to moan, so she quickly reassured him, "It's okay, you saved me, thank you," before realising one of its teeth was stuck in her skin.

Without thinking about it, she ripped it out of her cheek and squirted blood over the floor.

She pressed her hand against her face, trying to stop the bleeding. She could feel three cuts, each of them large, but it hadn't penetrated bone or muscle, and she had Boy to thank for that. It wasn't fatal, but she needed to ensure she didn't lose too much blood. She couldn't allow herself to feel faint, or pass out – she was going to need to be aware. There was a lot of fighting still to come.

As she knelt, applying pressure to her cheek, Boy knelt beside her and put his hand on hers. She forced a smile in appreciation.

"I'm sorry..." he said, and she wondered what for, then realised he thought he'd hurt her.

"It wasn't you, the knife didn't do this," she said. "It was the – the thing. The creature. Not you. Please don't worry."

She used her spare hand to stroke his cheek, and though he didn't look any happier, he seemed to look less sad.

She took her hand away from her cheek. She was just going to have to cope with blood trickling down her face. There were more important things to worry about.

Like how were they going to get out of this room?

They had to find a way. She had to get to the helm. She had to turn the ship around.

She rushed to the door and peered through the window. The corridor was full of them. Scuttling along the floor, crawling over the walls, hanging from the ceiling. She'd let three of them in and they'd only just survived. There was no way they could take on that many.

She turned her attention to Boy. They were trapped. They would die if they went into the corridor. Then she saw, beyond Boy, the land passing by the porthole, and she realised... there was another way out.

She huffed, and bemoaned her luck. She and Boy, who couldn't swim, were going to have to leave through the porthole, brave the ocean, and climb up the ladder instead.

And they were going to have to hurry, as the end of the land was coming into view.

CHAPTER FORTY-THREE

Paul was a goddamn action hero.

He was Bruce, he was Arnie, he was Chuck.

And he was positive that he looked just as cool, if not slicker.

The bag full of items was slung over his shoulder, the tank of propane in his right hand, and a bottle of alcohol in the other. He'd stuffed a lit piece of paper in the end of the bottle to create a home-made Molotov; so perfect he should have a show that teaches children how to try this at home.

He could set fire to anything he wished.

To that wall, to the deck, to himself.

It didn't matter that the corridors were made of steel. He had propane. He could light it all up if he felt like it.

Fire doesn't just give you power, it *is* power, and those that can wield it for their own purpose *become* power.

You are beautiful.

You are marvellous.

And, should you wish to be, you are God.

And he believed it.

Oh, how he believed it.

He felt it surging through him, every vein carrying the blood of the holy, every muscle twitching with excitement, every bone stiff with glory.

The creatures should worship him.

He was the ultimate predator, and it was time to get his prey.

He moved across the deck, imagining himself striding in slow motion, a coat billowing behind him, some out-dated heavy metal music confirming to an awed audience just how magnificent this man was.

They would all see it. All of them.

This new-found world was made for men like him.

He kicked open the door to find a corridor full of the little creatures. He threw a Molotov at them, setting a few alight.

They scattered, the heat of the flames too much for them, and he took his aerosol can and match and sprayed fire over any that didn't get away in time.

By the time he'd reached the fourth floor down, they had all returned to the safety of the ocean.

These were the most powerful creatures alive; the most ruthless, and most evil monsters ever to roam this earth – and he'd conquered those slaves of Hell.

Just like a god would.

Just him.

The one with all the power.

God.

"Where are you?" he bellowed. This time, he did not keep to a maniacal whisper; he roared, wanting not just to unsettle his opponent, but to terrify them.

As it was, he found them in the same room, trapped like ants in a jar.

Silly little girl. Creature of habit.

He rested his forehead against the window. Grinned. Watched as they backed away from him, pushing themselves against the far wall.

"Those nasty little fiends have gone now..." he sang. "You can come out..."

It occurred to him that he had keys.

Didn't he?

They were somewhere.

At least, he could find some.

But it wouldn't be anywhere near as fun.

"If you won't come out... I will make you come out..."

He doused a rag on propane. Lit it. Ducked beneath the window and held it at the base of the door. Its smoke spread beneath the crack. They coughed, and it delighted him so much he barely noticed that he was coughing too.

He stood, considering this to be a sufficient warning.

"Come out, or you'll have no choice but to come out..."

She turned away from him.

She turned away. From him.

She turned away.

From him.

His face curled, his fists clenched, his body tensed.

"How dare you..."

The cretin.

One does not simply turn away from God!

His lip curled and his mind raged and his body roared and he lifted his arm and struck the Molotov to the ground, creating a small fire to his left where the Molotov landed. He poured propane around it, and allowed the fire to spread.

He reached into his bag. Pulled out paper. Garlic. Oil. And he spread it across the crack beneath the door, forcing the flames to spread further.

But he left the corridor to his right alone, of course. He

poured no propane on the steel floor, meaning the fire did not spread. He was willing to grant the girl the opportunity of a safe passage to the deck. He wasn't a complete animal.

Paul turned back to the window, feeling his divine power surge through him.

"Come out, or I'll block your escape forever."

She looked back at him.

Shook her head.

And presented her middle finger.

Through his cracked lips he whispered, "Blasphemy…"

His anger overtook him. He knew what to do now. She'd made her decision. He wasn't planning on doing this, but now he had no choice.

He proceeded through the corridor, and went down the stairs, deeper and deeper, and further toward the rear of the boat.

Until he reached the engine room.

He had no idea what everything was. There were concealed boxes all labelled A_1, A_2, A_3, B_1, B_2, B_3 and so on. But these had to be the engines. They had to.

So, just to be sure, he spread propane across every single box. Then he lit his match and ran, through the corridor, to the stairs, to the deck, waiting for the explosion.

This ship would sink. He had seen to that.

CHAPTER FORTY-FOUR

There was no time left.

The land would soon disappear, the room would be filled with thick smoke clouds. She knew that, within minutes, she would not be able to see Boy any longer.

Then, just as she thought it could get any worse, the ship shook. But this wasn't the monster; this was a rumble from within. Something inside the ship had exploded.

What had Paul done?

She looked toward the porthole. It was beginning to submerge.

"Oh, God..." she gasped... *The ship is sinking.*

They had to get out.

She grabbed Boy's hand, looked into those eyes and, once again, requested the impossible from him: "I need you to trust me."

He gave her that look. The one that he used when he knew horror was about to come.

"We will die in this room, and there's only one way out."

He looked confused, then she turned her head toward

the window that led to the ocean, and he immediately shook his head.

She grabbed his head, kept him still, and forced him to look at her.

"If we leave through the corridor, Paul will kill us, or the fire will kill us, or those creatures will kill us. If we stay here, the smoke will kill us. We have no choice, Boy, so please, just... Trust me."

He held her gaze and, for a moment, looked like he was contemplating.

She didn't wait for a verdict.

She took the chair and swung it at the porthole.

It didn't make so much as a scratch.

In the distance, the land disappeared, and the distant sea took over the view once again.

No. Not now. Not ever. She was not going to give up her only chance of getting to land.

With a burst of adrenaline and determination she swung the chair again. No dent, no smash, no scratch.

She swung it again. And again. And again.

A small mark appeared, but that was it.

Her arms ached. She was exhausted.

She looked over her shoulder. Flames engulfed the corridor.

She lifted the chair once more and threw it. It did nothing.

She bent over. Hands on knees. Panting.

"I can't do it..."

She coughed. Smoke fogged the room. The surface of the water rose up the porthole. Death felt closer now than it ever had before.

She needed more attempts to smash that window, but her

muscles couldn't do much more. She had to, but they were too tired.

Boy took the chair from her and threw it at the window.

She lifted her head. Surprised. Pleasantly astonished that he had recognised her pain, and the situation.

Once again, he found a way to show her how very, very special he was.

He threw the chair again, and again.

But it wasn't enough. It wasn't smashing.

And they could only see water outside the porthole. The room had submerged.

So he climbed upon the bed and, instead of throwing the chair, he lunged it at the porthole without letting go, over and over, pounding it and pounding it until, finally, a crack ran through the middle.

She looked over her shoulder at the flames, then joined Boy on the bed. Once it smashed, they wouldn't have much time. They'd escape the fire, only for water to become their enemy.

Boy hit the window and cracked it further, then, with one large thrust, pounded the chair against the glass and smashed it completely.

"Well done!" she said as water poured in. She grabbed the porthole and tried to pull herself out, but the pressure of the water just shoved her back in.

Within seconds, the water's surface had reached the level of the bed. Cia knew it wouldn't be long before the room was completely submerged.

But what about the ship? Would it stay afloat, what with the fire and the flooding?

No time to think about that now...

She took hold of Boy's hand. The pressure of the water entering the room was too much. They would have to wait until the room was full until they could leave – which meant having to swim out. She'd need to help him.

"Hold your breath and follow me, okay?" she urged Boy.

He nodded.

She checked her knife was tucked into her belt. Took a second. Held her breath and, once the room was completely filled with water, she pulled herself through the porthole then turned to help Boy.

It didn't take long until they reached the ocean's surface, and they were floating, with her hands still gripping Boy.

She glanced over her shoulder and saw the faraway image of land. It was too far away to swim to, she needed the ship to get closer – for as long as it would last, at least; but it still provided her with the renewed determination she needed to get there.

"Come on," she told Boy, and they used the side of the ship to guide them toward the frontend.

CHAPTER FORTY-FIVE

Cia found the ladder they'd originally used to climb aboard, took hold of the first step, and dragged herself up, Boy following.

A low-pitched boom reverberated through the sea.

The creature was close.

She reached the railings and threw herself over, landing on her back, with Boy landing beside her. Her spine throbbed from the pain, but she ignored it, along with the throbbing of her cheek and the cuts stinging her hand.

She pushed herself to her feet and surveyed the catastrophe. The ship was sinking. Water rose in the distance, indicating that the creature was approaching. And Paul must be close. But she buried these thoughts somewhere in the back of her mind, took hold of Boy's hand, and ran to the cabin, finally reaching the helm.

She placed her hands on the wheel and pushed, putting all of her body behind it. It took a lot of force, but with Boy helping her, the ship began to turn – but not by much. Why wasn't it turning properly?

Did Paul destroy the engine room? Was that what the rumble was earlier?

Yet the ship was still turning a little, and without the engine it wouldn't turn at all... Meaning there must be more than one engine room, and either Paul didn't realise that or was too insane to care. Either way, the ship would still turn, she just needed to put more force into it.

She pushed, and pushed harder, using what little energy hadn't already been expended, and she could see Boy straining too. The front of the ship turned more and more toward the land, until it was perfectly aligned, and they were heading in the right direction.

She searched the controls for something else. Something to accelerate the ship. There was a lever, and she pushed it, and what little power the ship still had powered it forward, giving it that little extra speed.

Behind the land, dawn was becoming visible. The sun was small and gave little light, but it was beginning to rise, and the prospect of survival gave her hope – but only a little. They weren't there yet. But they were heading in the right direction. She just had to survive long enough to–

Fire flew past the window. It landed on the deck in front of the cabin and created a small fire, but did not spread across the steel deck.

She turned to look over her shoulder just in time to see Paul open the door to the cabin, a flaming bottle in his hand.

"No!" she screamed. He was about to destroy everything.

He threw the flaming bottle at her feet and she jumped out of the way. He smashed a bottle of something flammable on the floor, which allowed the fire spread quickly around the cabin, so she and Boy rushed out of the cabin and across the deck – but Paul left through the other door and blocked their escape.

His demented face was something to behold. He looked smug. His face twitched. His grin widened. Something in his eyes scared her more than she'd admit.

And the deck felt lower. Closer to the ocean surface. Cia peered at the land. The ship was not going to make it.

She glared at Paul.

"Why?" she asked, quietly at first, then again, louder. "Why!"

Paul shrugged. "Why not?"

"Do you want to die too?"

"Of course not; it is your sacrifice that allows me to live."

"What then? What will you live on with no ship?"

He shrugged again.

"What is wrong with you?" she said, her voice buckling under the pressure. "Are you just a maniac? Enjoying anarchy for anarchy's sake?"

Paul smiled at her and went to light another Molotov.

"Stop setting fire to things!" Cia said. "Can't you see what you're doing?"

But it only encouraged him, and he threw it across the deck. It landed inches away from Cia, and she rushed out of its way.

Boy hid behind her, finding his safe spot; but it wouldn't be safe for long.

She could see it behind Paul. A tank of propane.

He was going to set fire to the entire ship if she didn't stop him.

Enough.

Cia slid the knife out from her belt, felt her rage intensify her resolve, and ran toward him, screaming her war cry.

CHAPTER FORTY-SIX

Aw, how sweet.

The little girl wants to fight with the big boys.

Silly little cherub. Doesn't she realise, a mortal can't defeat a god?

She charged into his belly, and the surprise forced him to drop the propane.

He lifted himself up, lifted his fist, and sent it soaring into her cheek. She fell to the floor, leaving imprints of blood on his knuckles. At first, he felt smug that he managed to draw blood on his first strike, then he saw that she was already cut.

She used this opportunity, however, to take the tank of propane and drag it to the ship's edge.

"Don't you dare..." he warned her.

But she dared.

As he screamed, "*No!*" she threw it overboard.

"You... *bitch!*"

She snorted. "That's just a word used by simple men to describe a woman far too complex for their little minds."

What the fuck was she on about?

She was trying to throw him off. Challenging his

intellect. Defying his unworldly talents with her feeble words.

It didn't matter.

The ship was on sinking. She had nowhere to go. She would end up in the ocean, and the creature would have her, and he would be granted the grace of another few weeks of life.

He didn't think about what was next. Such logic escapes an irrational mind.

But what was that, in the distance?

Was that... land?

She'd turned the ship. He'd felt it move, and he'd thought it was just the destruction he'd created, but no...

She was trying to evade him.

Still.

Stupid little wretch. She should kneel down and obey him, not flee.

She charged at him again. This time, he simply swung his fist and landed it in the side of her head, knocking her onto her back.

She gazed upwards, her world going dizzy. He knew what it was like to take a punch, and seeing as she had a little girl's skull, it would hurt her a lot more than it hurt him. But still, the precious little thing, she tried to stand, tried to overcome the grogginess.

He simply lifted his boot and stamped its heel against her face.

Now all he needed to do was pick her up and throw her over the railings. He could place her in the fire, but the creature would most likely appreciate a meal that wasn't burnt.

He bent over to pick her up, but something else caught his attention.

The boy. He was running. At Paul.

Paul laughed.

The boy went to charge into him like Cia had, but Paul managed to catch the imbecile by the throat, which he squeezed, harder and harder, watching him choke.

The girl pushed herself up and sliced his kneecap with her knife, forcing him to fall to his knees and drop the boy.

"Just get somewhere safe!" she urged the boy, and he backed away.

Paul laughed. What a shit show of a man! Not only was he easily defeated, he also followed instructions from a girl.

He would *never* take instructions from a girl.

He pushed himself back to his feet, and the girl tried to do the same. Still, he was quicker, and he was afforded another stomp on her head, which forced her back to the floor.

A glance over the railings confirmed that the water's surface was rising. The explosions at the other end of the ship were growing more frequent, and it wouldn't be long until the ship either exploded or sank – and this girl would never reach her precious land.

She tried to get up again, but he stomped on her head again.

And, just as instructed, the boy didn't intervene.

This was a glorious moment of satisfaction. It was right up there with the murder of his wife, and the death of his wailing child. He'd squeezed the life out of them, and now he was going to remove any opportunity of survival from this girl who had caused him so much shit.

Enough was enough.

The time had come.

He was going to end this.

CHAPTER FORTY-SEVEN

THE WORLD SPUN.

Cia could smell nothing but smoke and sea. A strange combination. One that told her the situation was perilous but that, mercifully, she was still alive.

Paul stomped on her head again.

The ship lost focus for a moment.

Then she caught sight of Boy's face. Sitting where she'd told him to. Staying out of the way.

Not because he didn't want to intervene, but because he trusted her.

He knew she wouldn't submit.

He knew she had other plans.

She looked up at Paul. Saw a lifebuoy ring behind him.

Paul bent down, ready to pick her up, ready to throw her overboard and leave her to the creature.

She held the knife behind her back. Flexed her fingers over the handle.

He lifted his leg to kick her, but she swung the knife above her head and stuck it into his foot with all she had, pushing it through the vein and muscle and bone, then sticking the end into the lifebuoy ring.

He howled.

She leapt out of his reach. He swiped for her, but he couldn't move, thanks to the knife holding him in place.

She glanced over her shoulder.

Land. Coming closer.

The creature was still there, under the surface, but further out to sea. It wasn't approaching land. It was waiting for Paul to be done; for his offering to be prepared.

But his offering wasn't relenting.

Paul tried to remove the knife, but even just shifting caused him immense pain. He lifted his head back and cried out, leaving Cia the opportunity to dive forward and pull the knife out, forcing more howls of pain.

She didn't wait for a retaliation. She swung the knife into his stomach, pulled it out, and swung it again at his side, sticking it between his ribs.

This time, she left the knife in his chest, and stood back.

He lifted his head. Hunched over. Grabbing the knife. Blood trickling through the cracks between his fingers, down his arms, down his body.

She'd never seen a man look so insanely angry before.

Insane, yes. Angry, definitely. But maniacal rage...

His face twisted. His features distorted.

And all of this fury was aimed specifically in Cia's direction.

"You b–"

She didn't wait for another insult.

She charged forward, barged into his chest, and sent him stumbling backwards, into the railings. He toppled over and fell into the ocean with a satisfying splash.

She watched him for a moment. She knew she had other

things to attend to, but she wanted to relish his demise for as long as she could. His suffering. Ending the despair he'd caused her.

He tried to swim forward, but every movement created a shot of agony he couldn't seem to stand.

In the end, he was just bobbing in the ocean, screaming. As the ship moved away from him, she couldn't even tell what he was saying anymore. She could make out words like *bitch* and *whore* and *cunt*, but even after a while the obscenities faded, and he was just a figure floating amongst a vast blue sea.

He was her offering to the creature – one that would ensure her survival.

Boy tugged on her arm, forcing her to turn her attention back to the ship. Land was clear and visible under the morning sun, but the ship was a mess, and it was going under.

The ship would not make it to shore.

CHAPTER FORTY-EIGHT

No mercy.

That was Paul's mantra. The belief he lived his life by.

He hadn't given his wife any mercy when she'd begged for her life, appealing to his better nature, beseeching him to leave his poor family out of this. He'd gutted her and watched her bleed to death.

No mercy. *Ever.*

Except, as he floated in the sea, watching the ship grow smaller as it moved toward land, he felt like he could do with a little mercy.

He didn't feel so god-like anymore. He felt more like Noah and his ark – if the animals had been dead bodies, and Noah had been a sick bastard who'd fallen overboard.

Oh, how he couldn't help but chuckle. Floating in the ocean, helpless as a baby, thinking witty remarks as his resolve deteriorated...

He supposed he wouldn't be able to ask for more time.

Hell, he didn't even know what he'd do with the time if he was given it.

Swim to shore?

He could beg, but then again, he'd already done that. Every exchange he'd had with the creature had been on one knee, through his desperation not to be eaten.

His ego had buckled under the pressure.

Trevor Paul Scott knelt for no one.

Except, now he did. How foolish he felt.

No, to hell with that. It wasn't foolish. It was just the food chain. It made sense to understand where you were on it. The creature was above him, and he was above every other human.

No mercy.

Never any mercy.

Humans did not deserve it.

The waves grew more ferocious. He already knew it was there. It had been hovering by the ship throughout the whole ordeal, waiting for its offering. And now, it finally chose to reveal itself. Waterfalls cascaded off its domed head. Waves erupted under the pressure of its tentacles. It hovered, half above the surface, glaring at Paul.

Paul cackled. He couldn't help it. He'd seen the creature – a lot – but he'd never truly *looked* at it. And, honestly, it was a funny looking fucker.

"You look so stupid..." he said through snorts of hysteria.

The creature looked to the ship, going under as it approached land.

"Yeah, about that..."

What excuse could he give?

He'd been foolish. He'd underestimated them. He'd acted recklessly. He'd thought he could do better, he didn't really care, his heart wasn't in it blah blah blah yada yada yada.

Same old crap he'd use when his headmaster demanded to know why he'd punched another kid, or skived class, or told a teacher to shove their homework up their arse.

Only now, he'd finally met a foe he could be afraid of.

"I'll get them," he said.

He began to hatch his plan. How he could swim to shore, make it by dusk if he really tried, or maybe the creature could give him some help, could carry him on its stupid head or in a tentacle and place him on land so he could hunt for them.

But the creature wouldn't approach land. There were other creatures on land, and it wasn't where this one belonged.

He could find them and bring them back to sea then.

Yeah, he could do that.

He could maybe even find some more survivors – there were probably more in the buildings, hiding out – he could pretend he was good, like he was a friend, tell them there was somewhere across the sea they could be safe, and put them on a boat, a new one, he'd find one, and then bring them here and the creature could eat them and revel in its glory and it would all be just swell again.

Right?

"Just give me time and I'll–"

A tentacle swooped from under the surface, wrapping him up in its suckers, and lifting his helpless, flailing body into the air.

"What are you doing?" he screamed. "I'm on your side!"

But the creature was like him. A monster. It had no sides.

It only had people it devoured; those who lacked purpose.

It placed Paul in its mouth, stuck its teeth through his neck and torso, squeezing blood out of him like jam out of a doughnut.

It chewed, then swallowed, ending Paul's excruciating pain as its stomach digested his helpless body.

No mercy.

Never any mercy.

Ever.

CHAPTER FORTY-NINE

The sun reflected in the water, creating a path from ship to land.

But the ship was sinking, and it wasn't going to make it.

If only she could walk on water, she'd take the path illuminated by the sun and stroll neatly to refuge.

As it was, she was going to have to drag Boy through the water and do her best to keep him afloat.

"We're going to have to jump," she told Boy, gripping his arms. "Okay?"

He shook his head frantically.

But, as he looked around, Cia thought that surely even he must see that this ship would not keep them safe any longer. They had no choice but to abandon it.

But they were close enough to land that they could make it. She was sure of it. Just a little swimming and they'd be able to stand.

"We don't have a choice," Cia urged Boy, feeling tears in the corners of her eyes.

She wasn't sure why those tears were there. Perhaps it was the potential of relief, or it was the torment they had endured to get to this point, or perhaps it was despair at having to live in a world that constantly forced them to flirt with death.

But this was the choice in front of them. This was their fate. It was time to reach out of the pit of desolation and grip opportunity.

It was time to end this awful episode of their existence.

Another rumble of the ship almost knocked them over.

The second engine room. The fire from the first must have reached it.

At the other end of the ship, small explosions were growing into bigger explosions.

They didn't have time to deliberate.

"I will help you," Cia said. "I won't let you drown. But we have to jump. Now. Okay?"

Boy looked at the sinking ship, then back at Cia. He didn't nod, but he didn't shake his head either, and she took that as his reluctant affirmation.

"Let's go."

She grabbed his arm and pulled him to the railings. She climbed upon the top one, dragging Boy up, forcing him to do the same.

Then, hand in hand, they jumped, sinking into the water.

As they did, bright orange light shone above the water's surface, accompanied by the grumble of an explosion. Boy tried to reach the surface, but she held onto him and shook her head.

It wasn't time yet.

The orange light grew brighter, and pieces of debris shot into the water.

She grabbed Boy and dragged him away from the ship, remaining under the surface.

They both needed air. Her heart was racing and the adrenaline could only do so much.

But not yet.

The further they could get away, the less likely they were to be hit by flying objects – the ones that continued to surge through the water around them. Pieces of wood, of railing, of unidentifiable remains, all sinking past them.

Finally, she allowed them to rise to the surface, each of them taking in as much oxygen as they could.

She turned toward land, lit by the beaming sun in a clear, blue sky. Salvation welcomed them, and she was keen to greet it.

She held onto Boy with one hand and swam with the other. Her body felt heavy, but she forced herself to keep going, finding whatever energy she had left, constantly reminding herself – this was it. Just this final battle, this final surge.

Only, with every stroke, it became just that little bit tougher. Boy was struggling to stay afloat, and she was constantly having to pull him upwards, frequently checking that he was still with her

Land grew closer, and she saw sand. A beach. One she didn't recognise, but then again, all beaches looked the same.

Eventually, after a long arduous struggle, she was able to place her feet down, and she urged Boy to do the same.

With the water up to her mouth and to Boy's chest, they waded forward, pushing through the resistance, until the water came down to her neck, to her chest, then to her waist.

She felt like collapsing. She knew if she allowed herself, she'd quite easily pass out.

So she didn't allow herself.

She marched on through the water until it came up to her ankles. Her feet splashed in the shallow waves, with the only resistance coming from the wet clothes that clung to her body.

She collapsed on the beach. As did Boy. And they both gazed up at the sun, squinting. Her body felt dense. Hefty. Moving it would be like pulling a heavy weight.

But they couldn't stay here. She couldn't allow herself to fall unconscious somewhere so exposed.

So she sat up, and urged Boy to do the same.

"Come on," she said. "It's not over yet."

He moaned and refused, then a growl hung over the distance and he promptly sat up.

Funny, really. She was both scared and relieved to hear such a growl. Terrified that the creature might get her – but pleased that her enemy was finally one she was more familiar with.

"We just need to find somewhere to rest," she said. "Then we can stop. I promise."

She rolled onto her knees. Hesitated. Panted. Pushed herself up.

She shoved one leg forward, then the other.

Fed up of wet, heavy clothes, she removed her top and her trousers, discarding them, and Boy did the same. Once upon a time she'd have felt uncomfortable about being outside in her underwear. Now, however, it was the last thing she cared about.

When they eventually reached the nearby buildings, they found some clothes in a nearby shop. One with the window smashed and most of the clothes already looted – but there was a fleece and a pair of trousers. They were too big for her, but they would do. Boy managed to find t-shirt and shorts, and they carried on walking, and that's when Cia realised – this wasn't the UK.

The remains of the buildings didn't look like the

buildings where she'd grown up. They had a different style; she couldn't quite figure out what it was, but they did not look British.

Screw it. She could figure out where they were later. For now, they just needed to find somewhere to rest.

They found that place in a nearby building. One that used to be a block of flats. They chose a room that was quite a few floors up, despite how long it took them to climb the stairs – this was a survival technique she'd thought of long ago. If other survivors came in, they'd hear them ransacking the flats below, giving them enough time to avoid them.

She barged open the door to a flat on the third floor and went through the standard procedure of barricading themselves in with whatever they could find. Sofa. Television. Table.

Then they found the bed, fell upon it, and both drifted into a deep, satisfying sleep, their bodies finally able to rest.

TWO WEEKS LATER

CHAPTER FIFTY

Cia spent the next few weeks trying to figure out where they were. The buildings, although battered, broken and destroyed, were tall. They used a few of them to see what was around, climbing them in the morning and looking into the distance to see if there was a Thoral or a Lisker or a Waster nearby. One time, a Maskete had flown past, but they'd managed to hide before it caught sight of them.

She found a few newsagents and souvenir shops, and looked at the postcards – fortunately, people hadn't been that fussed about looting postcards and there were quite a few left – and she saw an image she recognised. A large tower. The Eiffel Tower.

But the Eiffel Tower was in Paris... Did this mean they were in Paris?

But they couldn't be in Paris as they'd arrived on a beach; Paris wasn't a coastal city, she sure of it. She looked over the other postcards, and found a picture of a beach just like the one they'd washed up on. In fact, there were quite a few postcards of this beach. Beneath some of them, the text stated clearly where they were: *Normandy*.

She remembered learning about World War Two in school. She remembered Churchill's famous speech about the D-Day battle that took place at Normandy – "We'll fight them on the beaches, we shall fight in the fields and in the streets, we shall fight in the hills." She remembered wanting to see this beach. To be where such a momentous moment of history occurred.

And she'd always wanted to see France.

She chuckled at the thought. This wasn't necessarily how she'd wanted to see it – she was thinking more of a holiday with her dad, rather than using its battered buildings for safety.

Then again, who would she rather go to France with than Boy?

He kept watch at the entrance. She approached him. Smiled. Put an arm around his shoulders.

Oh, the things she'd done for him. The people she'd killed, the times she'd risked her life, the monsters she'd faced.

And, truth was, she wouldn't do a single thing differently. Except, however, maybe not letting Paul out of his cell.

"I love you, kid," she said.

Boy just smiled and wrapped his arms around her.

"I love you too, Rosy."

She'd always wanted a brother. Now she had something better.

She had Boy.

Together, they left the souvenir shop, ready to search for food.

Then something caught her eye.

In another shop. A newsagent. Mostly looted. A few newspapers remained – all dated around the dates that the creatures had risen. And the picture was of a creature she'd hadn't seen before.

All the headlines were in French, and she didn't understand them. But, at the far side of the newspapers, was a newspaper in English. She supposed some French newsagents may have stocked some English newspapers.

The headline that drew her attention was:

Gigans Sighted in Paris.

What on earth was a Gigan?

She approached the article, and looked at another headline on the same page.

Army Warns Those in France to Stay Indoors: Striders Increasing in Quantity.

What on earth was a Strider?

And the image that went with it...

It was unlike any creature she'd ever seen.

She'd expected to see Thorals... Liskers... Masketes...

And it made her wonder...

Were the creatures indigenous? Had she only seen the creatures that had risen in her home country? The UK had Thorals, Liskers, and Masketes...

The English Channel had its own predators.

And France...

There was a picture of a Strider. It looked like a large insect, with long legs walking over buildings.

But there wasn't a picture of a Gigan. Just a news story:

> Sceintists have warned that the Gigan may be the deadliest creature discovered so far. Whilst other countries have had other creatures, such as Thorals and Masketes in the UK, France seems to be facing the most formidable opponent.

They have warned that the creature is lethal, and to even try to run from it would be pointless. Therefore, the French government has urged its citizens to hide. To take what canned goods they have and to await the response of the armed forces.

But the armed forces evidently hadn't managed to mount an attack, because, well... where were they? And what creature could be so bad that the government said that even trying to run from it was pointless?

How could it be so much worse than what she'd faced already?

A low-pitched sound reverberated in the distance, so grand in its volume Cia could feel it vibrating through her body.

She looked at Boy. He looked back at her, clueless, confused. It was a sound he evidently didn't recognise, and one she didn't either.

She peered into the distance, trying to see what it was.

Then again, as was warned by the fallen government, maybe she didn't want to see it. She grabbed Boy, took him into the nearest shop, into a door that led to a basement, and there they hid.

The sound came again, this time louder.

The ground shook under the thuds of its steps.

Jars on shelves rattled in the darkness.

The walls trembled.

She wished she was home.

She thought she'd faced the worst creature at sea. Yet, as she looked at the fear in Boy's face, she saw her own terror reflected. Their fight wasn't over.

They could hide here. They could ration what food

might have been left in this basement. Eventually, however, they would have to leave.

They would have to face the Gigan.

For now, she wouldn't think about tomorrow, or even that afternoon, or even the next five minutes. She just wrapped her arms around Boy, who did the same to her, and they held each other as the rattling and shaking and thudding faded into the distance.

They had each other.

Whatever the Gigan was, they could face it.

They had to. Her purpose was to keep Boy alive, and she had to face that task a day at a time.

She closed her eyes and rested her head against Boy's, and they stayed like this for some time.

Grateful for being alive, as they always were.

But fearful for what was to come, as they always were.

They would leave this room, hide, survive, flee, fight, run, be angry, be sad, be anything they had to be – there was more war to come if they were to endure the next month, or week, or even day.

For now, the Gigan could wait.

For now, they just had each other.

And that was how it was always going to be.

BOOK FIVE COMING SOON

JOIN RICK WOOD'S READER'S GROUP
FOR TWO FREE BOOKS

Sign up at www.rickwoodwriter.com/sign-up

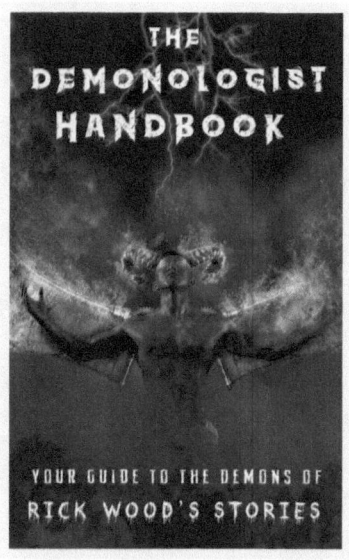

ALSO AVAILABLE BY RICK WOOD

BLOOD SPLATTER BOOKS

18+

PSYCHO B*TCHES

Rick Wood

www.ingramcontent.com/pod-product-compliance
Lightning Source LLC
LaVergne TN
LVHW041628060526
838200LV00040B/1483